Firefighting and Fire Safety Systems on Ships

This accessible reference introduces firefighting and fire safety systems on ships and is written in line with the IACS Classification Rules for Firefighting Systems. It covers the design, construction, use, and maintenance of firefighting and fire safety systems, with cross references to the American Bureau of Shipping rules and various Classification Society regulations which pertain to specific Classification Society rules. As such, this book:

- Focuses on basic principles in line with current practice.
- Is aimed at non-specialists.

The book suits professional seafarers, students, and cadets, as well as leisure sailors and professionals involved in the logistics industry. It is also particularly useful for naval architects, ship designers, and engineers who need to interpret the Class rules when developing shipboard firefighting systems.

T0407110

Firefighting and Fire Safety Systems on Ships

Alexander Arnfinn Olsen

CRC Press
Taylor & Francis Group
Boca Raton   London   New York

CRC Press is an imprint of the
Taylor & Francis Group, an **informa** business

Cover image: Ministry of Defence, licensed under the Open Government Licence v3.0
www.nationalarchives.gov.uk/doc/open-government-licence/version/3/

First published 2023
by Routledge
4 Park Square, Milton Park, Abingdon, Oxon OX14 4RN

and by Routledge
605 Third Avenue, New York, NY 10158

Routledge is an imprint of the Taylor & Francis Group, an informa business

British Library Cataloguing-in-Publication Data
A catalogue record for this book is available from the British Library

ISBN: 978-1-032-47302-4 (hbk)
ISBN: 978-1-032-47305-5 (pbk)
ISBN: 978-1-003-38552-3 (ebk)

DOI: 10.1201/9781003385523

Typeset in Sabon
by Apex CoVantage, LLC

Contents

PART 3
Additional requirements   **191**

9   Additional fire protection requirements   193

Preface

Classification Society[1] (Class) rules incorporate many of the requirements intended to prevent the onset of a fire onboard ships at sea. However, even with all the preventive measures taken, shipboard fires still occur with alarming frequency. Therefore, the proper design, installation, and operation of the vessel's firefighting systems are critical to the safety of a vessel and the personnel onboard. Since firefighting systems are so critical, the designs and arrangements of such systems should be carefully evaluated for compliance with Class requirements by the designer and Class engineering assessment staffs. This book has been developed to assist in a better understanding of the Class requirements for such systems. It is intended to provide a general overview of the generic Class requirements that should be considered during technical plan review activities. This book also discusses the basic scientific fundamentals of fire, as appropriate for a proper understanding and application of Class requirements. Accordingly, this book should be considered as general guidance only and the technical reviews of firefighting systems should verify compliance with all Class rules applicable to the specific vessel involved.

It should be noted that the scope of this book is limited to a review of Class requirements considered during the technical plan review of active firefighting systems onboard civilian merchant vessels. Passive fire protection arrangements, such as structural fire protection, as well as fire detection systems, are outside the scope of this book. Firefighting systems of offshore facilities and installations are also beyond the scope of this book. The review of firefighting systems for the International Maritime Organisation's (IMO's) *International Convention for the Safety of Life at Sea* (SOLAS) requirements fall out with the scope of this book as there are plenty of authoritative texts already available, including the *International Code for Fire Safety Systems* (FSS Code). However, in many cases, the Class rules for firefighting systems either incorporate or directly reference IMO SOLAS firefighting system requirements. Accordingly, within the discussions of the Class requirements for various firefighting systems, related 'interpretations' of the associated SOLAS requirements, as developed by the International

Association of Classification Societies (IACS), are provided. These IACS interpretations are called *Unified Interpretations* (UIs). As an IACS member, the Class is obligated to apply these UIs as appropriate interpretations of the SOLAS requirements, unless directed otherwise by Flag administrations. The interpretations provided in the UIs should be considered when conducting technical plan reviews. In addition to the IACS UIs, IMO MSC Circ.1120 'Unified Interpretations of SOLAS Chapter II-2, the FSS Code, the FTP Code and related Fire Test Procedures' (and its subsequent correction (Corr.1) and amendments per MSC.1 Circs.1436, 1491, and 1510) also provide guidance regarding SOLAS firefighting system requirements. Many of the MSC. Circ.1120 interpretations are based on the IACS UI interpretations, while others provide revised or additional interpretations of the SOLAS requirements. Several MSC.Circ. interpretations have been included in the SOLAS 2000 consolidated edition and subsequently in the SOLAS Consolidated Editions 2004, 2009, and 2014. Consequently, these requirements have also been incorporated into the Class rules. However, since the MSC Circulars interpretations identify useful information and guidance relative to certain Rule requirements, 'selected' IMO MSC interpretations have been incorporated where relevant.

Alexander Arnfinn Olsen

Southampton, November 2022

NOTE

1 Most Classification Societies are members of the International Association of Classification Societies (IACS), which includes the American Bureau of Shipping (www.eagle.org), Bureau Veritas (https://marine-offshore.bureauveritas.com/), China Classification Society (www.ccs.org.cn/ccswzen/), Croatian Register of Shipping (www.crs.hr/en-us/home.aspx), Det Norske Veritas (www.dnv.co.uk); Indian Registry of Shipping (www.irclass.org), Korean Register of Shipping; Lloyds Register (www.lr.org), ClassNK (www.classnk.or.jp); Russian Registry of Shipping (www.prs.pl), and RINA (www.rina.org).

Author's note

Throughout this book I have made use of the unilateral term, Class, as short-hand for classification society. Unless stated otherwise, please read 'Class' to mean all members of IACS, rather than as a reference to one specific classification society. The majority of Class rules and requirements are derived from the regulations promulgated by the IMO and its subcommittees. Other regulations are devised by national Flag State administrations, government authorities, and other parties with an interest in maritime safety. When the IMO, or its subcommittees, issues new guidance, these are readily adopted by classification societies as Class rules. Whilst each Class interprets this guidance according to their understanding, the end effect is the adoption of universally accepted, and enforced, rules. It is this universality that this book aims to provide.

Abbreviations

To simplify what is a complex subject area, the use of abbreviations has been kept to an absolute minimum, save for widely used and recognised terms, such as SOLAS and the IMO. For ease of use, the list below contains those terms which have been abbreviated, predominantly due to their continued use throughout this book.

ASTM	ASTM International, formerly the American Society for Testing and Materials
BLEVE	Boiling liquid expanding vapour explosion
BS EN	The list of BSI Standards includes BS standards of UK origin, BS EN standards published by CEN/CENELEC and BS ISO and BS IEC international standards.
BTU/lb	British thermal unit per pound[1]
Circs.	Circular
Class	Classification society (e.g., ABS, LR, RINA, DNV, and GL)
cm	Centimetre
CO	Carbon monoxide
CO_2	Carbon dioxide
DWT	Deadweight tonnage
EFPR	Emergency fire pump room
ESD	Emergency safety device
FSS	Code
ft	Foot
GPM	Gallons per minute
GRT	Gross registered tonnage

IACS	International Association of Classification Societies
IMDG	International Maritime Dangerous Goods (IMDG) Code
IMO	International Maritime Organisation
in	Inch
ISO	International Organisation for Standardisation
kPa	Kilopascal
l/min/m²	Litres/per minute/per metre squared
lb (lbs)	Pound/pounds
LFL	Lower flammable limit
LNG	Liquid natural gas
LPG	Liquid petroleum gas
m	Metre
MKS	International system of measurements (i.e. the metric system)
mm	Millimetre
MSC	Maritime Safety Council
NFPA	National Fire Protection Association (US)
pH	In chemistry, pH, historically denoting 'potential of hydrogen', is a scale used to specify the acidity or basicity of an aqueous solution. Acidic solutions are said to have lower pH values than basic or alkaline solutions.
psi	Per square inch
RORO	Roll on, roll off
SGR	Steering gear room
SI	International system of measurements (i.e. the metric system)
SOLAS	International Convention for the Safety of Life at Sea
UFL	Upper flammable limit
UI	Unified interpretation

NOTE

1 The British thermal unit is a unit of heat; it is defined as the amount of heat required to raise the temperature of one pound of water by 1 °F.

Part I

Science of combustion

Chapter 1

Chemistry of fire

Fire is a phenomenon with which we are all familiar. We use it in our daily lives to heat our homes and cook our meals. When harnessed, the power and energy from fire serve us well; however, when they are uncontrolled, a fire can quickly consume and destroy whatever lies in its path. While we are all familiar with fire, few of us are aware of its nature and complex processes. This chapter examines the phenomena and various mechanisms at work within a fire and is intended to provide a better understanding of the requirements in firefighting scenarios.

FUNDAMENTALS OF FIRE

Central to understanding the basic chemistry of fire is the process of oxidation, which is a chemical reaction between the molecules of a substance and the oxygen molecules in the surrounding atmosphere. There are many common examples of oxidation, including the rusting of iron, the tarnishing of silver, or the rotting of wood. What is known as *fire* is a chemical reaction involving the oxidation of the fuel molecules. However, the reaction occurs at a much faster rate and only under certain conditions (e.g. elevated temperatures and proper mixture). In addition, what is called *burning* or *combustion* is the continuous rapid oxidation of millions of fuel molecules. Recognising that the fire or combustion process is a chemical reaction (involving the oxidation of the fuel molecules) is critical to understanding the basics of the fire phenomena. The oxidation reaction is an exothermic process (i.e. one in which heat is given off). The molecules oxidise by breaking apart into individual atoms and recombine with the oxygen atoms to form new molecules. During this process, a certain amount of energy is released. In the examples of rusting iron or rotting wood, the amount of energy released is minimal since these oxidation processes occur at a very slow rate. However, in a fire, the oxidation rate of the fuel molecules is much faster. Because of this rapid reaction, energy is released at a much greater rate. The released energy is felt and seen in the form of heat and light. The more rapid the oxidation rate, the

DOI: 10.1201/9781003385523-2

greater intensity in which the energy is released. An explosion is, in fact, the oxidation of a combustible media at an extremely fast rate.

All substances exist in one of three states: as a solid, a liquid, or a vapour (gas). For the oxidation process to occur, there must be an adequate inter-mixing of the oxygen and fuel molecules. For fuel molecules in either a solid or liquid state, the molecules are tightly bound and cannot be effectively sur-rounded by the oxygen molecules in the atmosphere. Therefore, molecules in either a liquid or solid state are not directly involved in the rapid chemical reaction of oxidation in a fire. However, fuel molecules in a vapour state are free to mix with the atmosphere. These molecules become effectively surrounded by the oxygen molecules in the atmosphere and are available to become involved in the oxidation process. In fact, only fuel molecules in a vapour state are involved in the oxidation process. While fuel molecules in the solid or liquid states are not directly involved in the oxidation process, when heated, these molecules will move about more rapidly. If sufficient heat (energy) is applied, some fuel molecules break away from the surface to form a vapour just above the surface. This new vapour can now mix with oxygen and can become involved in the oxidation process. Accordingly, the fuel molecules in a solid or liquid state do serve as the source of additional fuel vapours when exposed to heat.

COMBUSTION PROCESS

The combustion process, or burning, is in fact the rapid oxidation of mil-lions of fuel molecules in the vapour form. Once there is sufficient oxygen and the fuel vapour molecules properly mix, an ignition source is typically needed for oxidation to be initiated. However, once oxidation is initiated, it is an exothermic process. If sufficient energy is released during the reac-tion to maintain the elevated temperature of surrounding oxygen and fuel molecules, and there are sufficient oxygen and vapourised fuel molecules available, then the oxidation process will continue. The heat released by the oxidation of the fuel molecules is radiant heat, which is pure energy, the same sort of energy radiated by the sun and felt as heat. It radiates, or travels, in all directions. Thus, part of it moves back to the seat of the fire, to the 'burn-ing' solid or liquid (the fuel). The heat that radiates back to the fuel is called radiation feedback. This part of the heat serves to release more vapours and serves to raise the vapour (fuel and oxygen molecule mixture) to the ignition temperature. At the same time, air is drawn into the area where the flames and vapour meet. The result is that the newly formed vapour begins to burn, and the flames increase, which starts a chain reaction. The burning vapour produces heat, which releases and ignites more vapour. The addi-tional vapour burns, producing more heat, which releases and ignites still more vapour. Provided there is sufficient fuel and oxygen available, the fire will continue to grow. For a fuel source with a limited amount of surface area available, the amount of vapour released from the fuel reaches a maximum rate and begins to level off, producing a steady rate of burning. This usually continues until most of the fuel has been consumed. When there is less fuel vapour available to oxidise, less heat is produced, and the process begins to die out. A solid fuel may leave an ash residue and continue to smoulder for some time, while a liquid fuel usually burns up completely.

Fire triangle

There are three components required for combustion to occur. These are:

- Fuel – to vapourise and burn.
- Oxygen – to combine with the fuel vapour.
- Heat – to raise the temperature of the fuel to its ignition temperature.

Together, these three components are called the *fire triangle*. There are two important factors to bear in mind when preventing and extinguishing a fire:

- If any of the three components are absent, then a fire cannot start.
- If any of the three components are removed, then the fire will go out.

It is important to have a clear understanding of these three components and their inter-reactions in a fire. The following sections will examine each of these components in further detail.

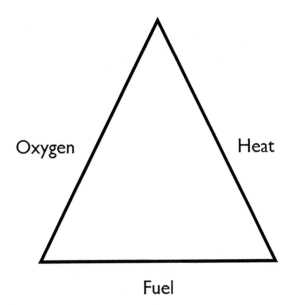

Figure 1.1 Fire triangle.

Fuel

Fuel is necessary to feed a fire, and without fuel, the combustion process will terminate. The fuel molecules involved in a fire must be in the vapour (gas) state. However, the initial fuel source may be in a solid, liquid, or gaseous state. Many examples of each type of these fuels can be found onboard a vessel. The following sub-paragraphs provide a brief discussion on the various types of fuels and some of the technical issues that impact their involvement in a fire.

Solid fuels

(1) Sources.

   The most obvious solid fuels are wood, cloth, and plastics. These types of fuels are found onboard a vessel in the form of cordage, dunnage, furniture, plywood, wiping rags, mattresses, and numerous other items. The paint used on bulkheads is considered a solid fuel. Vessels may also carry a variety of solid fuels as cargo, from baled materials to goods in cartons and loose materials, such as grain. Metals, such as magnesium, sodium, and titanium, are also solid fuels that may be carried as cargo.

(2) Pyrolysis.

   Before a solid fuel will burn, it must be changed to the vapour state. In a fire situation, this change usually results from the initial application of heat. The process is known as *pyrolysis*, which is generally defined as 'chemical decomposition by the action of heat'. In this case, the decomposition causes a change from the solid state to the vapour state. If the vapour mixes sufficiently with air and is heated to a high enough temperature (by a flame, spark, hot motor, etc.), then ignition results.

(3) Burn rate.

   The burning rate of a solid fuel depends on the rate at which vapours are generated, which depends on several criteria, including the configuration of the fuel surface. Solid fuels in the form of dust or shavings will burn much faster than bulky materials (i.e. small wood chips will burn faster than a solid wooden beam). Finely divided fuels have a much larger surface area exposed to the heat. Therefore, heat is absorbed much faster, vapourisation is more rapid and more vapour is available for ignition, allowing the fire to burn with great intensity and the fuel to be quickly consumed. A bulky fuel will burn longer than a finely divided fuel. Dust clouds are made up of very small particles. When a cloud of flammable dust (such as grain dust) is mixed well with air and ignited, the reaction is extremely quick, often with explosive force.

Such explosions have occurred on vessels during the loading and discharging of grains and other finely divided materials.

(4) Ignition temperature.

The ignition temperature of a substance (solid, liquid, or gas) is the lowest temperature at which sustained combustion will occur. Ignition temperatures vary among substances. The ignition temperature varies with bulk, surface area, and other factors. Generally accepted ignition temperatures of common combustible materials in various standardised configurations are provided in various handbooks.

Liquid fuels

(1) Sources.

The flammable liquids most found aboard a vessel are bunker fuel, lubricating oil, diesel oil, kerosene, oil-based paints, and their solvents. Cargoes may include flammable liquids and liquefied flammable gases.

(2) Vapourisation.

Flammable liquids release vapour in much the same way as solid fuels. The rate of vapour release is greater for liquids than for solids since liquids have less closely bonded molecules. In addition, liquids can release vapour over a wide temperature range. Gasoline starts to give off vapour at –43 °C (–45 °F). Since gasoline produces flammable vapour at normal atmospheric temperatures, it is a continuous fire risk, even without heating or back radiation. Heating increases the rate of vapour release and therefore the fire risk. Heavier flammable liquids, such as bunker oil and lubricating oil, release lesser amounts of vapours at atmospheric temperatures. However, the rate of vapourisation increases rapidly when heated. Some lubricating oils can ignite at 204 °C (400 °F). Because a fire reaches this temperature rapidly, oils that are directly exposed to a fire will soon become involved. Once a flammable liquid is burning, radiation feedback and the chain reaction of oxidation quickly increase flame production. The vapour produced by most flammable liquids is heavier than air. Vapour that is heavier than air is very dangerous because it will seek low places, dissipate slowly, and can quickly travel to a distant source of ignition. For example, vapour escaping from a container can travel along a deck and down deck openings until it contacts a source of ignition (such as a spark from an electric motor). If the vapour is properly mixed with air, then it will ignite and carry fire back to the leaky container, resulting in a fire.

(3) Burning.
Pound-for-pound, flammable liquids produce about 2.5 times more heat than wood when involved in a fire, and the heat is liberated three to ten times faster from liquids than from wood. These ratios illustrate quite clearly why flammable liquid vapour burns with such intensity. When flammable liquids spill, they expose a very large surface area and release a great amount of vapour; therefore, they can produce great amounts of heat when ignited. This is one reason why large open tank fires and liquid-spill fires burn so violently.

(4) Flash point.
The flash point of a liquid fuel is the lowest temperature at which the vapour pressure of the liquid is just sufficient to produce a flammable mixture at the lower limit of flammability. The flash points (temperatures) of liquids are determined in controlled tests and are usually reported as a 'closed cup' or an 'open cup' temperature. The flash point values of liquids are frequently established using the ASTM or *Pensky-Martens* testing apparatuses and procedures. However, other testing apparatuses are available. It is important to note that the typical value identified as the flash point of a flammable liquid is based upon atmospheric pressure. However, the value of the flash point for a particular liquid will vary as the atmospheric pressure to which the liquid is exposed increases or decreases.

Gaseous fuels

(1) Sources.
There are both natural and manufactured flammable gases. Those that may be found onboard a vessel include acetylene, propane, and butanes, as well as a number of liquefied gases carried as cargo on LNG and LPG vessels.

(2) Burning.
Gaseous fuels are already in the required vapour state. Only the proper intermixing with oxygen and sufficient heat are needed for ignition. Gases, like flammable liquids, do not smoulder. Radiation feedback is not necessary to vapourise the gas; however, some radiation feedback is still essential to the burning process (i.e. provide continuous re-ignition of the gas).

(3) Flammable range.
A flammable gas or the flammable vapour of a liquid must mix with air in the proper proportion to make an ignitable mixture. The smallest

percentage of a gas (or vapour) that will make an ignitable air-vapour mixture is called the Lower Flammable Limit (LFL) of the gas/vapour. If there is less gas in the mixture, it will be too lean to burn. The greatest percentage of a gas/vapour in an ignitable air-vapour mixture is called its Upper Flammable Limit (UFL). If a mixture contains more gas than the UFL, it is too rich to burn. The range of percentages between the LFL and UFL is called the flammable range of the gas or vapour. It is therefore important to realise that certain ranges of vapour-air mixtures can be ignited and to use caution when working with these fuels. The flammability ranges of specific types of fuels are published in various handbooks. Refer to such documentation for the product of concern.

Oxygen

Because the combustion process involves the oxidation of the fuel molecules, the availability of oxygen is vital for the process to exist. Accordingly, the second side of the fire triangle refers to the oxygen content in the surrounding air. Air normally contains about 21% oxygen, 78% nitrogen, and 1% other gases, principally argon, and therefore, sufficient oxygen is typically available unless some type of controlled atmosphere (i.e. inerted) is involved.

Heat

For fuel molecules to undergo the oxidation process and result in a self-supporting fire, the molecules must be at elevated temperatures (i.e. ignition temperature). Without this elevated temperature, there will be no rapid oxidation or combustion of the fuel molecules. Further, the generation of additional fuel vapours is largely dependent upon feedback radiant heating of the fuel, except for gaseous fuels. Therefore, heat is the third side of the fire triangle. The production of energy from the initial reaction tends to raise the temperature of other molecules to the necessary elevated temperatures and tends to create the self-supporting nature of fire.

Fire tetrahedron

The fire triangle (Figure 1.1) is a simple means of illustrating the three components required for the existence of fire. However, it lacks one important element when trying to understand the nature of fire and the effectiveness of extinguishing mechanisms available. It does not consider the *chemical reaction* of the oxidation process and the chain reaction needed for a fire to continue to exist. The fire tetrahedron (Figure 1.2) provides a better representation of the combustion process. A tetrahedron is a solid figure with

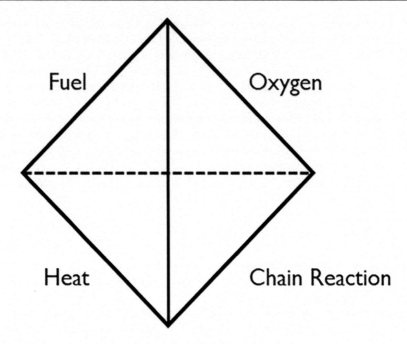

Figure 1.2 Fire tetrahedron.

four triangular faces and is useful for illustrating the combustion process because it shows the chain reaction, and each face touches the other three faces. The basic difference between the fire triangle and the fire tetrahedron is that the tetrahedron illustrates how flaming combustion is supported and sustained through the chain reaction of the oxidation process. In a sense, the chain reaction face keeps the other three faces from falling apart. This is an important point, because the extinguishing agents used in many modern portable fire extinguishers, automatic extinguishing systems, and explosion suppression systems directly attack and break the chain reaction sequence to extinguish a fire.

Extinguishing effect considering the fire tetrahedron

A fire can be extinguished if any one side of the fire tetrahedron can be destroyed. If the fuel, oxygen, or heat is removed, the fire will die out. Likewise, if the chain reaction is broken, the resulting reduction in vapour generation and heat production will also result in the extinguishing of the fire. (However, additional cooling with water may be necessary where smouldering or re-flash is a possibility.)

Removing the fuel

One way to remove fuel from a fire is to physically move it. In most instances, this is an impractical firefighting technique. However, it is often possible to move nearby fuels away from the immediate vicinity of a fire so that the fire does not extend to these fuels. Sometimes the supply of liquid or gaseous fuel can be cut off from a fire. When a fire is being fed by a leaky heavy fuel oil or diesel line, it can be extinguished by closing the proper valve. If a pump is supplying liquid fuel to a fire in the engine room, the pump can be shut down to remove the fuel source and thereby extinguish the fire; these arrangements are required by the Class rules.

Removing the oxygen

A fire can be extinguished by removing its oxygen or by reducing the oxygen level in the atmosphere. Many extinguishing agents (e.g., carbon dioxide and foam) extinguish a fire with smothering action by depriving the fire of oxygen. This extinguishing method is difficult (but not impossible) to use in an open area. Gaseous smothering agents, such as carbon dioxide, are blown away from an open deck area, especially if the vessel is underway. However, a fire in a galley garbage container can be snuffed out by placing a cover tightly over the container, blocking the flow of air to the fire. As the fire consumes the oxygen in the container, it becomes starved for oxygen and is extinguished. Tank vessels that carry petroleum products are protected by foam systems with monitor nozzles on deck. The discharged foam provides a barrier over the fuel, and when used quickly and efficiently, a foam system can extinguish a sizeable deck fire. To extinguish a fire in an enclosed space such as a compartment, engine room, or cargo hold, the space can be flooded with carbon dioxide. When the carbon dioxide enters the space and mixes with the atmosphere, the percentage of oxygen in the atmosphere is reduced and extinguishing results. This medium is used to combat fires in cargo holds. However, it is important to seal the enclosed space as reasonably gas tight as possible to maintain the concentration of CO_2 and the reduction in oxygen. It is also important to note that there are some cargoes known as oxidising substances that release oxygen when they are heated or, in some instances, when they encounter water. Substances of this nature include the hypochlorites, chlorates, perchlorates, nitrates, chromates, oxides, and peroxides. All contain oxygen atoms that are loosely bonded into their molecular structure, that is, they carry their own supply of oxygen, enough to support combustion. This oxygen is released when the substances break down, as in a fire. For this reason, burning oxidisers cannot be extinguished by removing their oxygen. Instead, for most oxidisers, large amounts of water, limited by vessel stability safety needs, are used to accomplish extinguishing effect. Oxidisers are hazardous materials and, as such, are regulated by the Flag administration in association with the SOLAS regulations.

Eliminating the heat

The most used method of extinguishing fire is to remove the source of heat. The base of the fire is attacked with water to cool the fuel surface, which reduces the amount of fuel vapour being generated. Water is a very effective heat absorber. When properly applied, it absorbs heat from the fuel and absorbs much of the radiation feedback. As a result, the chain reaction is indirectly attacked both on the fuel surface and at the flames. The production of vapour and radiant heat is reduced. Continued application will control and extinguish the fire. Heat, which is a critical element in the fire tetrahedron, can be transferred from a fire by one or more of three methods: conduction, radiation, and convection. Each of these methods of heat transfer must be considered when engaging a fire.

(1) Conduction.
Conduction is the transfer of heat through a solid body. For example, on a hot stove, heat is conducted through the pot to its contents. Wood is ordinarily a poor conductor of heat, but most metals are good conductors. Since most vessels are constructed of metal, heat transfer by conduction is a very real potential hazard. Fire can and will move from one hold to another, one deck to another, and one compartment to another via heat conduction through the steel structure, unless structural fire protection arrangements are provided to prevent such propagation.

(2) Radiation.
Heat radiation is the transfer of heat from a source across an intervening space. No material substance is involved. The heat travels outward from the fire in the same manner as light, that is, in straight lines. When it contacts a body, it is absorbed, reflected, or re-transmitted. Absorbed heat increases the temperature of the absorbing body. For example, radiant heat that is absorbed by an overhead will increase the temperature of that overhead, perhaps enough to ignite its paint. Radiant heat travels in all directions, unless it is blocked, and can extend a fire by heating combustible substances in its path, causing them to produce vapour and then igniting the vapour.

(3) Convection.
Convection is the transfer of heat through the motion of heated matter (e.g. through the motion of smoke, hot air, heated gases produced by the fire, and flying embers). When it is confined as within a vessel, convective heat moves in predictable patterns. The fire produces lighter-than-air gases that rise towards high parts of the vessel. Heated air, which is lighter than cool air, also rises, as does the smoke produced by combustion. As these heated combustion products rise, cool air takes

their place. The cool air is heated in turn and then rises to the highest point it can reach. As the hot air and gases rise from the fire, they begin to cool, and as they do, they drop down to be reheated and rise again. This is the convection cycle. It is important to recognise that heat originating from a fire on a lower deck will travel horizontally along passageways and then upward via ladder and hatch openings, and it can ignite flammable materials in its path.

Breaking the chain reaction

Another mechanism for destroying the fire tetrahedron, and therefore, extinguishing the fire, is by the interruption of the chain reaction. Once the chain reaction sequence is broken, the heat generation is reduced. This reduces the fuel vapour generation, as well as the heating of the vapour oxygen mixture, and as a result, the fire is extinguished. The extinguishing agents commonly used to attack the chain reaction and inhibit combustion are dry chemicals and Halon alternatives. These agents directly attack the molecular structure of compounds formed during the chain reaction sequence by scavenging the 'O' and 'OH' radicals. The breakdown of these compounds adversely affects the flame-producing capability of the fire. It should be borne in mind that these agents do not cool a deep-seated fire or a liquid whose container has been heated above the liquid's ignition temperature. In these cases, the extinguishing agent must be maintained on the fire until the fuel has cooled down naturally.

Hazardous and combustible materials

There are several hazards produced by a fire, including flames, heat, gases, and smoke. Each of these combustion products can cause serious injuries or death and should be considered in the overall scope of firefighting arrangements onboard a vessel.

Flame

The flaming region of a fire is that portion of the combustion zone where the fuel and oxygen molecules are of an appropriate mixture and temperature to support the oxidation process. The direct contact with flames can result in totally or partially disabling skin burns and serious damage to the respiratory tract. To prevent skin burns during a fire attack, crewmen must maintain a safe distance from the fire unless they are properly protected and equipped for the attack. This is the reason that protective clothing, such as fireman's outfits, is required by Class rules. Respiratory tract damage can be prevented by wearing breathing apparatus.

Heat

As a result of a fire, the temperature in an enclosed space can reach temperatures more than 93 °C (200 °F) very rapidly, and the temperature can build up to over 427 °C (800 °F) quickly. Temperatures above 50 °C (122 °F) are hazardous to humans, even if they are wearing protective clothing and breathing apparatus. The dangerous effects of heat range from minor injury to death. Direct exposure to heated air may cause dehydration, heat exhaustion, burns, and blockage of the respiratory tract by fluids. Heat also causes an increased heart rate. A firefighter exposed to excessive heat over an extended period could develop hyperthermia, a dangerously high fever that can damage the nerve centre.

Combustion gases

The gases produced by a fire depend mainly on the fuel. The most common hazardous gases are carbon dioxide (CO_2), the product of complete combustion, and carbon monoxide (CO), the product of incomplete combustion. CO is the more dangerous of the two. When air mixed with CO is inhaled, the blood absorbs the CO before it will absorb oxygen. The result is an oxygen deficiency in the brain and body. Exposure to a 1.3% concentration of CO will cause unconsciousness in two or three breaths and death in a few minutes. Carbon dioxide works on the respiratory system. Above normal CO_2 concentrations in the air reduce the amount of oxygen that is absorbed in the lungs. The body responds with rapid and deep breathing, which is a signal that the respiratory system is not receiving sufficient oxygen. When the oxygen content of air drops from its normal level of 21% to about 15%, human muscular control is reduced. At 10% to 14% oxygen in air, judgement is impaired and fatigue sets in. Unconsciousness usually results from oxygen concentrations below 10%. During periods of exertion, such as firefighting operations, the body requires more oxygen, and these symptoms may then appear at higher oxygen percentages. Depending on the fuel source, there may be several other gases generated by a fire that are of equal concern to firefighters. Therefore, anyone entering a fire must wear an appropriate breathing apparatus.

Fire vapours (smoke)

Fire vapours, or smoke as they are commonly but incorrectly referred to as, are a visible product of fire that adds to the problem of breathing. It is made up of carbon and other unburned substances in the form of suspended particles. It also carries the vapours of water, acids, and other chemicals, which can be poisonous or irritating when inhaled.

In this chapter, we have explored the underpinning principles of what fire is, how it exists, and more importantly, what we need to do to put a fire out. In the next chapter, we will turn our attention to the different classifications of fire – that is, the main source of fuel for each type of fire – and the implications for crew members onboard ship when tackling each classification or fire type.

Chapter 2

Classifications of fire

The characteristics of fires and the effectiveness of extinguishing agents differ with the fuels involved. While some extinguishing agents are very effective on fires involving certain fuels, they may be much less effective or even hazardous for use on other types of fires. Take, for example, the use of a portable water extinguisher. Water is a good extinguishing medium and is very effective on deep-seated fires, such as burning wood or rubbish. However, a firefighter would not want to use a portable water extinguisher on a fire involving a 'live' electrical panel or switchboard due to the conductivity of the water and the possible shock that could result. Considering the different types of fuels that may be involved in a fire, the different types of extinguishing agents available and the different mechanisms which the various agents use to extinguish a fire, it is important to be able to identify the type of fire on which a particular medium will be effective. The job of selecting the proper extinguishing agent has been made easier by the classification of fires into five types, or classes, lettered 'A' through 'D' and 'F' or 'K', based upon the fuels involved. Within each class are fires involving those materials with similar burning properties and requiring similar extinguishing agents. Thus, knowledge of these classes is essential to efficient firefighting operations, as well as familiarity with the burning characteristics of materials that may be found aboard a vessel.

The IMO mentions two standards in IMO Resolution A.951(23) which define the various classes of fires. The first is the International Standards Organisation (ISO) Standard 3941, and the second is the National Fire Protection Agency (NFPA) 10 Standard. Table 2.1 identifies these classes of fire as they are listed in IMO Resolution A.951(23). IMO Resolution A.951(23) is included in Annex of the *International code for fire safety system* (FSS Code). While the types of combustibles covered by ISO and NFPA are very similar for Classes 'A', 'B', and 'D', the combustibles covered by Class 'C' designation differ substantially. Considering that the classification of the various types of combustibles used in the Class rules (e.g., the ABS Class Notes in 4–7–3/Table 4 of the *Rules for building and classing steel vessels* and 4–5–1/Table 1 of the *Rules for building and classing steel vessels under*

DOI: 10.1201/9781003385523-3

Table 2.1 Fire Classifications

ISO Standard 3941	NFPA 10
Class A: Fires involving solid materials, usually of an organic nature, in which combustion normally takes place with the formation of glowing embers.	Class A: Fires in ordinary combustible materials, such as wood, cloth, paper, rubber, and many plastics.
Class B: Fires involving liquids or liquefiable solids.	Class B: Fires in flammable liquids, oils, greases, tars, oil base paints, lacquers, and flammable gases.
Class C: Fires involving gases.	Class C: Fires, which involve energised electrical equipment where the electrical non-conductivity of the extinguishing medium is of importance. (When electrical equipment is de-energised, extinguishers for class A or B fires may be used safely.)
Class D: Fires involving metals.	Class D: Fires in combustible metals such as magnesium, titanium, zirconium, sodium, lithium, and potassium.
Class F: Fires involving cooking oils.	Class K: Fires involving cooking grease, fats, and oils.

90 m (295 ft) in length) more closely follow the NFPA designations, we will refer to the NFPA classifications. In the remainder of this chapter, the fuels, their burning characteristics, by-products, etc., within each fire class are discussed in more detail.

CLASS 'A' FIRES

Class 'A' fires involve three groups of materials commonly found onboard a vessel, including:

- Wood and wood-based materials.
- Textiles and fibres.
- Plastics and rubber.

The following sections discuss Class 'A' fires involving each of these materials.

Wood and wood-based materials

Wood products are often involved in fire, mainly because of their many uses. Marine uses include furniture, furnishings, dunnage, and staging, as well as numerous other uses. Wood-based materials are those that contain processed wood or wood fibres and include some types of plywood and panelling, paper, cardboard, and press board. The burning characteristics of wood and wood-based materials depend on the wood involved. For example, seasoned, air-dried maple (a hardwood) produces greater heat upon burning than does pine (a softwood) that has been seasoned and dried similarly. However, all these materials are combustible, and they will char, smoulder, ignite, and burn under certain conditions. Wood is composed mainly of carbon, hydrogen, and oxygen, with smaller amounts of nitrogen and other elements. In the dry state, most of its weight is in the cellulose. Some other ingredients found in dry wood are sugars, resins, gums, esters of alcohol, and mineral matter.

Burning characteristics of wood and wood-based materials

The ignition temperature of wood depends on many factors, such as size, shape, moisture content, and type. Generally, the ignition temperature of wood is taken to be about 204 °C (400 °F). However, it is believed that if wood is subjected to temperatures above 100 °C (212 °F) over a long period, under certain conditions, ignition can take place. Similarly, the rate of combustion and heat release rate of wood and wood-based materials depend heavily on the physical form of the material, the amount of air available, the moisture content, and other such factors. For wood to become involved in a fire, the solid components of the surface of the wood must first be heated to the point where pyrolysis, the process whereby the solid components on the surface are converted to combustible vapours, is sufficient to support combustion. The heat necessary to produce pyrolysis can come from several sources, including direct contact with a flame, contact with some other heated element, or resulting from the radiant heat of a separate fire. The combustible vapours resulting from the pyrolysis are released from the surface of the wood and mix with the surrounding air. When the mixture of combustible vapour and air is within the flammable range, any source of ignition may ignite the combustible vapour mass almost instantly. Even without an ignition source, if the surface temperatures rise sufficiently, autoignition can occur. Flames move across the surface of combustible solids in a process called flame spread. Flame spread is the result of adjacent surfaces being heated by the existing flames to a point where the adjacent surface produces sufficient flammable vapours to support and fuel combustion. It is important to note that the orientation of the adjacent surface to the fire does play a role in the rate of flame spread. Flames will typically spread faster in an upward direction since such locations are heated by radiant heat from the flame, as well as convective heat from the fire plume. The process of flame spread will continue until all fuel is consumed or there is not sufficient heat available to promote adequate pyrolysis of adjacent surfaces. Bulky solids with a small surface area (e.g., a heavy wood beam) burn more slowly than solids with a larger surface area (e.g., a sheet of plywood), and solids in chip, shaving, or dust form (wood, metal shavings, sawdust, grains, and pulverised coal) burn even more rapidly, since they represent a much larger total area per mass of fuel. The larger surface area allows combustible vapour to be generated and released at a greater rate. (This is also true of flammable liquids. A shallow liquid spill with a large area will burn off more rapidly than the same volume of liquid in a deep tank with a small surface area.)

By-products of combustion of wood and wood-based materials

Burning wood and wood-based materials produce water vapour, heat, carbon dioxide, and carbon monoxide. The reduced oxygen levels and the carbon monoxide present the primary hazard to crew members and firefighters. In addition, wood and wood-based materials produce a wide range of aldehydes, acids, and other gases when they burn. By themselves or in combination with the water vapour, these substances can cause irritation at least, and in high enough concentrations, most produced gases are toxic.

Textiles and fibres

Textiles, in the form of clothing, furniture, carpets, canvas, burlap, ropes, and bedding, are used extensively in the marine environment, and others are carried as cargo. In addition, almost all textile fibres are combustible. These two facts explain the frequency of textile-related fires and the many deaths and injuries that result.

Natural textiles

(a) Plant fibres.
Vegetable fibres consist largely of cellulose. They include cotton, jute, hemp, flax, and sisal. Cotton and the other plant fibres are combustible [the ignition temperature of cotton fibre is around 400 °C (752 °F)]. Burning vegetable fibres produce heat and smoke, carbon dioxide, carbon monoxide, and water. The ease of ignition, rate of flame spread, and amount of heat produced depend on the construction and finish of the textile and on the design of the finished product.

(b) Organic fibres.
Organic fibres, such as wool and silk, are solid and are chemically different from vegetable fibres. They do not burn as freely, and they tend to smoulder. For example, wool is a protein. It is more difficult

to ignite wool than cotton. Given the ignition temperature of wool fibre is around 600 °C (1112 °F), it burns more slowly and is easier to extinguish.

Synthetic textiles

Synthetic textiles are fabrics woven wholly or mainly of synthetic fibres. Such fibres include rayon, acetate, nylon, polyester, acrylic, and plastic wrap. The fire hazards involved with synthetic textiles are sometimes difficult to evaluate, owing to the tendency of some of them to shrink, melt, or drip when heated. Rayon and acetate resemble plant fibres chemically, whereas most other synthetic fibres do not. Almost all are combustible in varying degrees but differ in ignition temperature, burning rate, and other combustion features.

Burning characteristics of textiles and fibres

Many variables affect the way in which a textile burns. The most important are the chemical composition of the textile fibre, the finish on the fabric, the fabric weight, the tightness of weave, and any flame-retardant treatment. Vegetable fibres ignite easily and burn readily, giving off large amounts of heavy smoke. Partially burned vegetable fibres may present a fire risk, even after they have been extinguished. Half-burned fibres should always be removed from the fire area to a location where re-ignition of the material would not create an additional problem. Most baled vegetable fibres absorb water readily. The bales will swell and increase in weight when large quantities of water are used to extinguish fires in which they are involved. Wool is difficult to ignite, and it tends to smoulder and char rather than to burn freely, unless it is subjected to considerable external heat. However, it will contribute towards a fierce fire. Wool can absorb a large amount of water – a fact that must be considered during prolonged firefighting operations. Silk is a less dangerous fibre. It is difficult to ignite, and it burns sluggishly. Combustion usually must be supported by an external source of heat. Once set on fire, silk retains heat longer than any other fibre. In addition, it can absorb a great amount of water. Spontaneous ignition is possible with wet silk. There may be no external evidence that a bale of silk had ignited, until the fire burns through to the outside. The burning characteristics of synthetic fibres vary according to the materials used, but the characteristics of some of the more common synthetics are given in Table 2.2.

These characteristics are based on small-scale tests and may be misleading. Some synthetic fabrics appear to be flame-retardant when tested with a small flame source, such as a match. However, when the same fabrics are subjected

Table 2.2 Burning Characteristics of Synthetic Fibres

Synthetic fibre	Burning characteristics
Acetate	Burns and melts ahead of the flame. Ignition point is like cotton.
Acrylic	Burns and melts. Ignition temperature approx. 560 °C (1040 °F). Softens at 235–330 °C (455–626 °F).
Nylon	Supports combustion with difficulty. Melts and drips; melting point, 160–260 °C (320–500 °F). Ignition temperature approx. 425 °C (797 °F) and above.
Polyester	Burns readily. Ignition temperature 450–485 °C (842–905 °F). Softens at 256–292 °C (493–558 °F) and drips.
Plastic wrap	Melys.
Viscose	Burns similarly to cotton.

to a larger flame or a full-scale test, they may burst into flames and burn completely while generating quantities of black smoke.

By-products of combustion of textiles and fibres

All burning materials produce hot gases (called fire gases), flame, heat, and smoke, resulting in decreased oxygen levels. The predominant fire gases are carbon monoxide, carbon dioxide, and water vapour. Burning vegetable fibres such as cotton, jute, flax, hemp, and sisal give off large amounts of dense smoke. Jute smoke is particularly acrid. Burning wool gives off dense, greyish-brown smoke. Another product of the combustion of wool is hydrogen cyanide, a highly toxic gas. Charring wool forms a sticky, black, tar-like substance. Burning silk produces a large amount of spongy charcoal mixed with ash, which will continue to glow or burn only in a strong draught. It emits quantities of thin grey smoke, somewhat acrid in character. Silk may produce hydrogen cyanide gas under certain burning conditions.

Plastics and rubber

A variety of organic substances are used in manufacturing plastics. These include phenol, cresol, benzene, methyl alcohol, ammonia, formaldehyde, urea, and acetylene. The cellulose-based plastics are largely composed of cotton products. However, wood flour, wood pulp, paper, and cloth also play a large part in the manufacturing of many types of plastic. Natural rubber is obtained from rubber latex, which is the juice of the rubber tree. It is combined with such substances as carbon black, oils, and sulphur to make commercial rubber. Synthetic rubbers are like natural rubber in certain characteristics. Acrylic, butadiene, and neoprene rubbers are some of the synthetic types.

Burning characteristics of plastics and rubber

The burning characteristics of plastics vary widely and depend upon the specific material involved, as well as the form of the product (solid sections, films and sheets, foams, moulded shapes, synthetic fibres, pellets, or powders). Most major plastic materials are combustible at least to some extent, and in a major fire, all contribute fuel to the fire. Plastics may be divided roughly into three groups with respect to burning rates:

(1) Materials that either will not burn at all or will cease to burn if the source of ignition is removed.
    This group includes asbestos-filled phenolics, some polyvinyl chlorides, nylon, and the fluorocarbons.
(2) Materials that are combustible, burn relatively slowly, but may or may not cease to burn when the source of ignition is removed.
    These plastics include the wood-filled formaldehydes (urea or phenol) and some vinyl derivatives.
(3) Materials that burn without difficulty and can continue to burn after the source of ignition is removed.

Included in this group are polystyrene, the acrylics, some cellulose acetates, and polyethylene.

In a class of its own is the oldest well-known form of plastic, celluloid, or cellulose nitrate plastic. It is typically considered the most dangerous of the plastics. Celluloid decomposes at temperatures of 121 °C (250 °F) and above with great rapidity, and without the addition of oxygen from the air. Flammable vapour is produced by the decomposition. If this vapour is allowed to accumulate and is then ignited, it can explode violently. It will burn vigorously and is difficult to extinguish. The caloric value of rubber is roughly twice that of other common combustible materials. For example, rubber has a heating value of 17.9×106 kJ (17,000 BTU/lb), whereas pine wood has a value of 8.6×106 kJ (8,200 BTU/lb). Most types of natural rubber soften when burning and may thus contribute to rapid fire spread. Natural rubber decomposes slowly when first heated. At about 232 °C (450 °F), it begins to decompose rapidly, giving off gaseous products that may result in an explosion. The ignition temperature of these gases is approximately 260 °C (500 °F). Synthetic rubbers behave similarly, though the temperature at which decomposition becomes rapid may be somewhat higher. This temperature ranges upward from 349 °C (660 °F) for most synthetics, depending on the ingredients. Latex is a water-based emulsion and *so does not present fire hazard.*

By-products of combustion of plastics and rubber

Burning plastic and rubber produce the fire gases, heat, flame, and smoke. These materials may also contain chemicals that yield additional combustion products of a toxic or lethal nature. The type and amount of smoke generated by a burning plastic material depend on the nature of the plastic, the additives present, whether the fire is flaming or smouldering, and what ventilation is available. Most plastics decompose when heated, yielding dense to very dense smoke. Ventilation tends to clear the smoke, but usually not enough for good visibility. Those plastics that burn cleanly yield less dense smoke under conditions of heat and flame. When exposed to flaming or non-flaming heat, urethane foams generally yield dense smoke, and in almost all cases, visibility is lost quickly. Hydrogen chloride is a product of combustion of chlorine-containing plastics, such as polyvinyl chloride, a plastic used for insulating certain electrical wiring. Hydrogen chloride is a deadly gas that has a pungent and irritating odour. Burning rubber produces dense, black, oily smoke that has some toxic qualities. Two of the noxious gases produced in the combustion of rubber are hydrogen sulphide and sulphur dioxide. Both are dangerous and can be lethal under certain conditions.

Locations of class 'A' materials on shipboard settings

Although vessels are constructed of metal and may appear incombustible, there are many flammable products aboard. Practically every type of material (Class 'A' and otherwise) may be carried as cargo. It may be in the cargo holds or on deck, stowed in containers or in bulk stowage. In addition, Class 'A' materials are used for many purposes throughout the vessel. In accordance with SOLAS regulation II-2/5.3.2, a certain quantity of combustible materials may be used in the construction of facings, mouldings, decorations, and veneers in accommodation and service spaces. In addition, the furnishings found in passenger, crew, and officer accommodations are usually made of Class 'A' materials. Lounges and recreation rooms may contain couches, chairs, tables, bars, television sets, books, and other items that may be constructed of Class 'A' materials. Other areas in which Class 'A' materials may be located include the following:

- Bridge contains wooden desks, charts, almanacs, and other such combustibles.
- Wood in many forms may be found in the carpenter shop.

- Various types of cordage are stowed in the boatswain's locker.
- Emergency locker on the bridge wing contains rockets and/or explosives for the line-throwing gun.
- Undersides of metal cargo containers are usually constructed of wood or wood-based materials.
- Lumber for dunnage, staging, and other uses may be stored below decks.
- Large numbers of filled laundry bags are sometimes left in passageways, awaiting movement to and from the laundry room.

Extinguishing effect of class 'A' fires

It is a fortunate coincidence that the materials most often involved in fire, Class 'A' materials, may best be extinguished by the most available extinguishing agent, water, provided by the required fire main system. In addition, other types of extinguishing mediums, such as foam and certain types of dry chemicals, are also effective on Class 'A' combustibles.

CLASS 'B' FIRES

Class 'B' fires involve two groups of materials commonly found onboard vessels. These are:

- Flammable liquids.
- Flammable gases.

Flammable liquids

Class regulations refer to a flammable fluid or liquid as any fluid, regardless of its flash point, liable to support flame, and that 'aviation fuel, diesel fuel, heavy fuel oil, lubricating oil, and hydraulic oil are all to be considered flammable fluids'. Accordingly, all types of liquids that will burn are typically considered to be flammable fluids, including oil-based paints and solvents.

Burning characteristics of flammable liquids

As previously noted, it is the vapour of a flammable liquid, rather than the liquid itself, which burns or explodes when mixed with air and ignited. These liquids will vapourise when exposed to air and at an increased rate when heated. Flammable vapour explosions most frequently occur within a confined space, such as a tank, room, or structure. The violence of a flammable vapour explosion depends upon:

(1)   Concentration and nature of the vapour.
(2)   Quantity of vapour-air mixture present.
(3)   Type of enclosure in which the mixture is confined.

The flashpoint is commonly accepted as the most important factor in determining the relative hazard of a flammable liquid. However, it is not the only factor. The ignition temperature, flammable range, rate of evaporation, reactivity when contaminated or exposed to heat, density, and rate of diffusion of the vapour also determine how dangerous the liquid is. However, once a flammable liquid has been burning for a short time, these factors have little effect on its burning characteristics. The burning rates of flammable liquids vary somewhat, as do their rates of flame travel. The burning rate of gasoline is 15.2 to 30.5 cm (6 to 12 in) of depth per hour, and for kerosene, the rate is 12.7 to 20.3 cm (5 to 8 in) of depth per hour. For example, a pool of gasoline 1.27 cm (1/2 in) deep could be expected to burn itself out in 2.5 to five minutes.

By-products of combustion of flammable liquids

In addition to the usual combustion products, there are some products that are peculiar to flammable liquids. Liquid hydrocarbons normally burn with an orange flame and give off dense clouds of black smoke. Alcohols normally burn with a clean blue flame and very little smoke. Certain terpenes and ethers burn with considerable boiling of the liquid surface and are difficult to extinguish. Acrolein (acrylic aldehyde) is a highly irritating and toxic gas produced during the combustion of petroleum products, fats, oils, and many other common materials. Liquid paint can also burn fiercely and gives off much heavy black smoke. Explosions are another hazard of liquid paint fires. Since paint is normally stored in tightly sealed cans or drums up to 150–190 l (40–50 gal) capacity, fire in any paint storage area may easily heat up the drums and cause them to burst due to excessive pressure. The contents are likely to ignite very quickly, possibly with explosive force with the exposure to air.

Locations of flammable liquids on shipboard settings

Large quantities of flammable liquids, in the form of heavy fuel oil, diesel oil, lubricating oil, and hydraulic oil, are also stowed aboard a vessel, for use in propelling and generating electricity. In addition, heavy fuel oil and diesel oil are frequently heated during the purification process or in preparation for injection into a boiler or an engine, which introduces additional hazards due to the additional volume of vapours capable of being created and the existence of those vapours at elevated temperatures. Fires involving these Class 'B' combustibles are usually associated with machinery spaces. In addition, flammable liquids of all types are carried as cargo by tank vessels, and flammable liquids in smaller packages may be found in holds of vessels carrying dangerous goods, as well as in the tanks of vehicles being transported in ferries or RORO vessels. Most paints, varnishes, lacquers, and enamels, except those with a water base, present a high fire risk in storage or in use. Accordingly, vessels are required by Class to carry such products in designated paint or flammable liquid lockers. Other locations where combustible liquids may be found include the galley (hot cooking oils) and the various mechanics shops and spaces where lubricating oils are used. Fuel and diesel oil may also be found as residues and films on and under oil burners and equipment in the engine room.

Extinguishing effect of flammable liquids

In general, Class rules require a fixed firefighting system to be installed in spaces specifically subject to Class 'B' fires. Various types of fixed systems are discussed in detail in following sections and are in addition to the required fire main system, as well as the various required portable and semi-portable appliances. A few examples are provided as follows:

- *Machinery spaces.* Category 'A' machinery spaces are required to be fitted with a fixed gas extinguishing system, fixed water spray system, or high expansion foam system.
- *Cargo tanks.* In addition to an inert gas system intended to maintain the cargo tanks in an inerted condition, oil and chemical carriers with cargo tanks carrying flammable liquid cargoes are required to be fitted with a foam system for coverage on the deck. The cargo pump rooms for such spaces are also required to be fitted with fixed fire extinguishing systems.
- *RORO spaces.* RORO spaces are required to be fitted with either a fixed gas extinguishing system or a water spray system.
- *Paint and flammable liquid lockers.* Paint and flammable liquid lockers must be fitted with a CO^2 system, a fixed water spray system, or a dry powder system. For paint lockers and flammable liquid lockers of floor area less than 4 m^2 (43 ft^2) having no access to accommodation spaces, portable fire extinguisher(s), which can be discharged through a port in the boundary of the lockers, may be Class approved.

Flammable gases

While most solids and liquids can be vapourised when their temperature is increased sufficiently, the term 'gas' is taken to mean a substance that is in the gaseous state at atmospheric temperatures and pressures. In the gaseous state, the molecules of a substance are not held together, but are free to move about and take the shape of its container. Any gas that will burn in the normal concentrations of oxygen in air is considered a flammable gas. As with other gases or vapours, a flammable gas will burn only when its concentration in air is within its combustible range and the mixture is heated to its ignition temperature. Flammable gases are usually stored and transported aboard vessels in one of three ways:

- *Compressed.* A compressed gas is one that, at normal temperatures, is entirely in the gaseous state under pressure in its container.
- *Liquefied.* A liquefied flammable gas is one that, at 37.8 °C (100 °F), has a Reid vapour pressure of at least 2.8 bar (40 psi). At normal temperatures, it is partly in the liquid state and partly in the gaseous state under pressure in its container.
- *Cryogenic.* A cryogenic gas is one that is liquefied in its container at a temperature far below normal temperatures and at low to moderate pressures.

Basic hazards

The hazards presented by a gas that is confined in a container differ from those presented when a gas that has escaped from its container. They will be discussed separately, although these hazards may be present simultaneously in a single incident.

Hazards of confinement

When the exterior of a tank or cylinder containing gas is exposed to a fire, the internal pressure of the container increases due to the heating of the gas. If the pressure increases sufficiently, a gas leak or a container failure could result. In addition to the increasing pressure, heating of the container surface (e.g., contact of flames with the container and radiant heat) can reduce the strength of the container material. To prevent the failure of tanks or cylinders containing compressed flammable gases, pressure relief valves and fusible plugs are typically installed. When the pressure exceeds a set limit, the relief valve opens allowing gas to flow out of the container, normally through a relief piping system to some safe location, thereby reducing the internal pressure. Typically, a spring-loaded device closes the valve when the pressure is reduced to a safe level. A fusible plug is another protective device. The fusible plug is constructed of a metal that will melt at a specific temperature. The plug seals an opening in the body of the container, usually near the top. Heat from a fire, threatening the tank or cylinder, causes the metal plug to melt allowing the gas to escape through the opening. Excessive pressure within the tank is prevented. However, the opening cannot be closed, and the gas will continue to escape until the container is empty. Failure of the containment can occur when these safety devices are not installed or fail to operate properly. Another cause of failure is the very rapid build-up of pressure in a container at a rate that the pressure cannot be relieved through the safety valve opening fast enough to prevent the excessive build-up of pressure.

Container protection
Compressed or liquefied gas represents a great deal of energy being held within the container. If the container fails, this energy is released – often very rapidly and violently. The gas escapes and the container or container pieces are scattered. Failure of a liquefied flammable gas container due to exposure to a fire is of great concern due to the concern with boiling liquid-expanding vapour explosion, or BLEVE (pronounced 'blevey'). Because of these concerns, the Class rules require relief valves to be of a capacity that can handle the flow rate of vapours during a maximum fire exposure.

(1) Hazards of gases released from confinement
    The hazards of a flammable gas that has been released from its container depend on the properties of the gas, as well as the location and conditions where it is released. Any released flammable gas can present a danger of a fire or explosion or both. A volume of released gas will not burn or explode if either:
    • The concentration of gas is not sufficient to result in a flammable air-gas mixture.
    • Any resulting flammable air-gas mixture does not encounter an ignition source.

A volume of released gas will burn without exploding if:
    • There is an insufficient concentration of gas anywhere but in the immediate vicinity of the point of release due to the dissipation of the gas into the atmosphere.
    • Because the initial release of gas was ignited so quickly that sufficient volumes of the unignited air-gas mixture did not have time to accumulate. These types of fires are normally referred to as jet fires.
      Where a sufficient volume of flammable gas is released in a manner that creates an unignited gas-air mixture within the flammable concentration range, an explosion can, and typically will, occur, if ignited. Such concentrations can be readily achieved if a flammable gas is released in an enclosed area. However, even on an open deck, if a massive release occurs, a gas cloud of sufficient concentration and volume can be created (unless ignition

at the point of release occurs immediately). Explosions of such open area gas clouds are known as open-air explosions or space explosions. Liquefied gases, hydrogen, and ethylene are subject to open-air explosions.

Properties of some common gases

Properties of several flammable gases are presented in the following pages. These properties lead to varying degrees and combinations of hazards when the gases are confined or released.

(1) *Acetylene*. Acetylene is composed of carbon and hydrogen. It is used primarily in chemical processing and as a fuel for oxyacetylene cutting and welding equipment. It is non-toxic and has been used as an anaesthetic. Pure acetylene is odourless, but the acetylene in general use has an odour due to minor impurities mixed in with the gas. Acetylene is subject to explosion and fire when released from its container. It is easier to ignite than most flammable gases and it burns more rapidly. This increases the severity of explosions and the difficulty of venting to prevent explosion. Acetylene is only slightly lighter than air, which means it will mix well with air upon leaving its container.

(2) *Anhydrous ammonia*. Anhydrous ammonia is composed of nitrogen and hydrogen. It is used primarily as fertiliser, as a refrigerant, and as a source of hydrogen for the special atmospheres needed to heat-treat metals. It is a relatively toxic gas, but its sharp odour and irritating properties serve as warnings. However, large clouds of anhydrous ammonia, produced by large liquid leaks, have trapped and killed people before they could evacuate the area. Anhydrous ammonia is subject to explosion and fire (and presents a toxicity hazard) when released from its container. However, its high LEL and low heat of combustion tend to minimise these hazards. In unusually tight locations, such as refrigerated process of storage areas, the release of the liquid or a large quantity of gas can result in an explosion.

(3) *Ethylene*. Ethylene is composed of carbon and hydrogen and is used in chemical processing, such as the manufacture of polyethylene plastic. Smaller amounts are used to ripen fruit. It has a wide flammable range and burns quickly. While non-toxic, ethylene is an anaesthetic and asphyxiant, and is subject to explosion and fire when released from its container. Ethylene has a wide flammable range and high burning rate. In several cases involving rather large outdoor releases, open-air explosions have occurred.

(4) *Liquefied Natural Gas (LNG)*. LNG is a mixture of materials, all composed of carbon and hydrogen. The principal component is methane, with smaller amounts of ethane, propane, and butane. LNG is non-toxic but is an asphyxiant. It is used as a fuel. LNG is shipped as a

cryogenic gas in insulated tanks. LNG is subject to explosion and fire when released from its container.

(5) *Liquefied Petroleum Gas (LPG)*. LPG is a mixture of materials, all composed of carbon and hydrogen. Commercial LPG is mostly either propane or normal butane, or a mixture of these with small amounts of other gases. It is non-toxic but is an asphyxiant. It is used principally as a fuel and, in domestic and recreational applications, sometimes known as 'bottled gas'. LPG is subject to explosion and fire when released from its container. As most LPG is used indoors, explosions are more frequent than fires. The explosion hazard is accentuated by the fact that 3.8 l (1 gal) of liquid propane or butane produce 75 to 84 m^3 (245 to 275 ft^3) of gas. Large releases of liquid-phase LPG outdoors have led to open-air explosions.

Many of the liquefied flammable gases, such as LPG and LNG, are transported in bulk on gas carriers, and fire protection requirements for such vessels are provided in the IMO gas code. Additionally, acetylene is typically found stored in cylinders for use onboard.

Extinguishing flammable gases

Flammable gas fires can be extinguished with dry chemicals that interrupt the chain reaction of the combustion (oxidation) process. The installation of a fixed dry chemical system is required on gas carriers. In addition, water spray systems are required to provide heat shields for certain portions of the structure from the radiant heat of gas fires. The procedure for fighting a flammable gas fire tends to involve allowing the gas to burn until the flow can be shut off at the source. Extinguishing effect is normally not attempted unless such extinguishing effect leads to shutting off the fuel flow, since the release of unignited gas can result in a more dangerous explosion. Until the flow of gas supplying the fire has been stopped, firefighting efforts are normally directed towards protecting areas that may contain materials that may be ignited by flames or radiated heat from the fire. Water, in the form of straight streams and fog patterns, is usually used to protect such exposures. When the gas is no longer escaping from its container, the gas flames should go out. However, if the fire is extinguished before shutting off the gas flow, firefighters must be careful to prevent the ignition of gas that is being released.

CLASS 'C' FIRES

Electrical equipment involved in fire, or in the vicinity of a fire, may cause electric shock or burns to firefighters. This section discusses some electrical installations found aboard a vessel, their hazards, and the extinguishing effect of fires involving electrical equipment.

Types of equipment

The main electrical installations onboard a vessel which are most at risk of fire include:

- *Generators.* Generators are machines that produce electrical power. These machines are usually driven by internal combustion engines, or in older vessels, steam produced in an oil-fired boiler. The electrical wiring in the generator may be insulated with a combustible material. Any fire involving the generator, or its prime mover, will involve a high risk of electrical shock to personnel.
- *Panel boards.* A panel board has fuses and automatic devices for the control and protection of lighting and power circuits. The switches, fuses, circuit breakers, and terminals within a panel board all have electrical contacts. These contacts may develop considerable heat, causing dangerously high temperatures and unnecessary operation of over current devices, unless they are maintained in good condition. Over current devices are provided for the protection of conductors and electrical equipment and open the circuit if the current in that circuit produces an excessively high temperature.
- *Switches.* Switches are required for the control of lights and appliances and for disconnecting motors and their controllers. They are also used to isolate high-voltage circuit breakers for maintenance operations. Switches may be of either the air-break or the oil-break type. In the oil-break type, the device that interrupts the circuit is immersed in oil. The chief hazard is the arcing produced when the switch is opened. In this regard, oil-break switches are the more hazardous of the two types. The hazard increases when a switch is operated much beyond its rated capacity, when its oil is in poor condition or when the oil level is low. Then the arc may vapourise the remaining oil, rupture the case, and cause a fire. However, if properly used and maintained, these switches present no hazard.
- *Electric motors.* Many fires are caused by electric motors. Sparks or arc, from short circuiting motor windings or improperly operating brushes, may ignite the motor insulation or nearby combustible material. Other causes of fires in motors include overheating of bearings due to poor lubrication and grimy insulation on conductors preventing the normal dissipation of heat.

Electrical faults that cause electrical fires

The main types of electrical faults that cause fires are:

- *Short circuits.* If the insulation separating two electrical conductors breaks down, a short circuit occurs. Instead of following its normal path, the current flows from one conductor to the other. Because the electrical resistance is low, a heavy current flow causes intense local heating. The conductors become overloaded electrically, and they may become dangerously overheated unless the circuit is broken. If the fuse or circuit breaker fails to operate, or is unduly delayed, fire can result and spread to nearby combustible material.

- *Overloading of conductors.* When too large an electrical load is placed on a circuit, an excessive number of current flows and the wiring overheats. The temperature may become high enough to ignite the insulation. The fuses and circuit breakers that are installed in electric circuits will prevent this condition. However, if these safety devices are not maintained properly, their failure may result in a fire.
- *Arcing.* An arc is pure electricity jumping across a gap in a circuit. The gap may be caused intentionally (as by opening a switch) or accidentally (as when a contact at a terminal becomes loose). In either case, there is intense heating at the arc. The electrical strength of the arc and amount of heat produced depend on the current and voltage carried by the circuit. The temperature may easily be high enough to ignite any combustible material near the arc, including insulation. The arc may also fuse the metal of the conductor. Then, hot sparks and hot metal may be thrown about and set fire to other combustibles.

Hazards of electrical fires

Although there are myriad hazards associated with electrical fires, the two primary hazards are:

- *Electric shock.* Electric shock may result from contact with live electrical circuits. It is not necessary to touch one of the conductors of a circuit to receive a shock. Any conducting material that is electrified through contact with a live circuit will suffice. Thus, firefighters are endangered in two ways. First, they may touch a live conductor or some other electrified object while groping about in the dark or in smoke. Second, a stream of water or foam can conduct electricity to firefighters from live electrical equipment. Moreover, when firefighters are standing in water, both the chances of electric shock and the severity of shocks are greatly increased.
- *Burns.* Many of the injuries suffered during electrical fires are due to burns alone. Burns may result from direct contact with hot conductors or equipment, or from sparks thrown off by these devices. Electric arcs can also cause burns. Even persons at a distance from the arc may receive eye burns.

Locations of electrical equipment on shipboard settings

Electric power is essential to the operation of a vessel. The equipment that generates, controls, and delivers this power is found throughout the vessel. Some of this equipment, such as lighting devices, switches, and wiring, is common and easily recognised.

- *Engine room.* The source of the vessel's electric power is its generators. At least two generators are situated in the engine room and often three or four generators are fitted. While the others are operating, one is always held in reserve for backup duty. The generators supply power to the main electrical switchboard, which is typically in the same area as the generators in the engine room. The switchboard houses the generator control panels, as well as a distribution section. If a fire breaks out in the vicinity of the generator switches or the main switchboard, the vessel's engineer can stop the generator by mechanical means. This will de-energise the panel board and switches. Also nearby is the engine room console, which contains controls for the fire pumps, ventilating fans, engineer's signal alarm panel, temperature detection system, and other engine room equipment.
- *Emergency generator room.* An emergency generator and switchboard are available for use on most vessels in case the main generator fails. It will provide power for emergency lighting and equipment only. They are situated in the emergency generator room, which is outside the main engine room.
- *Passageways.* Electrical distribution and lighting panel boards are located along passageway bulkheads. Much of the vessel's electrical wiring is placed in the passageway overheads. Access panels are provided in these overheads to allow work on the wiring, and these panels can be removed to check the area for fire extension.
- *Other locations.* The bridge contains much electrical equipment, including the radar apparatus, bridge console, smoke detector indicating panel, and lighting panel boards. Below decks, in the bow and stern, are electrical control panels for the capstan and winch motors. A power panel board in the machine shop controls the electric arc welding machine, buffer and grinder, and drill press and lathe. There is still much more electrical equipment located throughout every vessel. The important point is that the hazards of live electrical equipment must be considered whenever a shipboard fire is being fought.

Extinguishing electrical fires

When any type of electrical equipment is involved with fire, its circuit should be de-energised. However, recognising that the circuit may not be de-energised, the fire must be extinguished using a non-conducting agent, such as

CO_2 or dry chemical. In considering the application of a fire extinguishing agent, an electrical circuit or panel should always be considered energised.

CLASS 'D' FIRES

Metals are commonly considered to be non-flammable. However, they can contribute to fires and fire hazards in several ways. Sparks from the ferrous metals, iron and steel, can ignite nearby combustible materials. Finely divided metals are easily ignited at high temperatures. Various metals, especially in finely divided form, are subject to self-heating under certain conditions, resulting in fires. Alkali metals such as sodium, potassium, and lithium react violently with water, liberating hydrogen, and sufficient heat is generated in the process to ignite the hydrogen. Most metals in powder form can be ignited as a dust cloud, and violent explosions have resulted. In addition to all this, metals can injure firefighters through burning, structural collapse, and toxic fumes. Many metals, such as cadmium, give off noxious gases when subjected to the high temperatures of a fire. Some metallic vapours are more toxic than others. However, breathing apparatus should be used whenever fires involving metals are fought.

Hazards and characteristics of specific metals

The hazards and characteristics of specific metals which crew members need to be aware of are:

- *Aluminium.* Aluminium is a light metal with good electrical conductivity. In its usual forms, it does not present a problem in most fires. However, its melting point of 660 °C (1,220 °F) is low enough to cause the collapse of unprotected aluminium structural members. It is for this reason that aluminium is not permitted to be used in a fire main, even on an aluminium vessel. Aluminium chips and shavings have been involved in fire, and aluminium dust is a severe explosion hazard. Aluminium does not ignite spontaneously and is not considered to be toxic. In addition, if aluminium is mixed with or smeared into iron oxide (rust) and then struck, a spark can be generated which can ignite an existing flammable vapour. This is the reason that the Class rules prohibit the use of aluminium paint within hazardous areas, such as those on oil carriers.
- *Iron and steel.* Iron and steel are not considered combustible. In the larger forms, such as structural steel, they do not burn in ordinary fires. However, fine steel wool or dust may be ignited, and iron dust is a fire and explosion hazard when exposed to heat or flame. Iron melts around 1,535 °C (2,795 °F) and ordinary structural steel around 1,430 °C (2,606 °F).

- *Magnesium*. Magnesium is a brilliant white metal that is soft, ductile, and malleable. It is used as a base metal in light alloys for strength and toughness. Its melting point is 649 °C (1,200 °F). Dust or flakes of magnesium are easily ignited, but in solid form it must be heated above its melting point before it will burn. It then burns fiercely with a brilliant white light. When heated, it reacts violently with water and all moisture.
- *Titanium*. Titanium is a strong white metal, lighter than steel, which melts at 2,000 °C (3,632 °F). It is mixed with steel in alloys to give high working temperatures. It is easily ignited in smaller forms (titanium dust is very explosive), though larger pieces offer little fire hazard. Titanium is not considered toxic.

Locations of class 'D' metals on shipboard settings

The metal principally used in the construction of vessels is steel. However, aluminium, its alloys, and other lighter metals are used to build the super-structures of some vessels. The advantage of aluminium lies in the reduction of weight. A disadvantage, from the firefighting viewpoint, is the compara-tively low melting point of aluminium as compared to that of steel. In addi-tion to the material used for the vessel itself, metals are carried in most forms as cargo. Generally, there are no stowage restrictions regarding metals in solid form. On the other hand, the metallic powders of titanium, alumin-ium, and magnesium must be kept in dry, segregated areas. The same rules apply to the metals potassium and sodium. It should be noted here that the large containers used for shipping cargo are usually made of aluminium. The metal shells of these containers have melted and split under fire conditions, exposing their contents to the fire.

Extinguishing class 'D' fires

Metals that form a fire risk are usually carried as cargoes and fall under the category of dangerous goods. Frequently, there is a violent reaction with water, which may result in the spreading of the fire and/or explosion. No one method of extinguishing effect has proven effective for all fires involving metals. Sand, graphite, various other powder extinguishing agents, and salts of different types have been applied to metallic fires with varying success. Recognising the unique fire risks that individual dangerous goods present, the International Maritime Organisation has established emergency procedures for such products. These emergency procedures typically identify the extin-guishing agent to be used in the case of a fire involving these products and are found in appendices to the IMDG Code (*International maritime danger-ous goods code*), as well as the IMSBC Code (*International maritime solid bulk cargoes code and supplement*). In reviewing the firefighting arrangements for a vessel intended to carry cargoes considered to be a Class 'D' fire hazards, the aforementioned emergency procedures should be consulted.

CLASS 'F' OR 'K' FIRES

Class 'F' or 'K' fires involve unsaturated cooking oils in well-insulated cooking appliances. Fires that involve cooking grease, fats, and oils are designated Class 'F' under the European/Australasian systems and 'Class K' under the American system. Though such fires are technically a subclass of the flammable liquid/gas class, the special characteristics of these types of fires, namely, the higher flashpoint, are considered important enough to recognise them separately. Wet chemical can be used to extinguish these types of fires.

Hazards and characteristics of fires caused by cooking greases, fats, and oils

To meet the demand for faster cooking, a new generation of well-insulated, high-efficiency cooking appliances was introduced and is widely used in ship galleys nowadays. These appliances heat faster, cook hotter, and cool slower. They are also more economical to operate since less energy is needed. To accommodate this new technology, higher temperature vegetable cooking oils were needed. Deep fat fryers require high-temperature vegetable cooking oils. Health concerns have also contributed to reduced use of animal fats or lard as a cooking oil. This has fire protection consequences. The fat/oil used in deep fat fryers is normally heated to about 182 °C (360 °F) and the auto-ignition temperature of most animal fat is in the range of 288–316 °C (550–600 °F) compared to 363 °C (685 °F) or higher for vegetable oils. After the fat/oil has been in use for several days, the auto-ignition temperature is lowered by 28 °C (50 °F) or more. The older the grease, the greater the hazard. When a temperature controller malfunctions, the temperature of the fat/oil being heated can readily increase to the auto-ignition point and the vapours will catch fire from the excess temperature. All the fat/oil in the fryer, not just the surface, will be at the auto-ignition point and then fire occurs. The fire may also ignite the grease in the range overhead hood, allowing the fire to spread into the grease-coated ductwork leading up to the exhaust outlet. The heat emanating from the ductwork may ignite structure members composed of combustible materials in the galley and other spaces where the ductwork passes through. Burning continues until the temperature of the fat/oil is lowered below the flash point (or the fuel is exhausted). Insulated cooking appliances tend to retain heat and to slow cooling. This makes a fire more difficult to extinguish. Extinguishing agents used in galley fire protection systems typically have an alkaline base. When the agent is discharged on a fire, a process called *saponification* occurs. The alkaline base of the extinguishing agent combines with the fatty acids in the cooking oil and forms a soap-like substance which blankets the surface. This helps smother the fire. However, saponification occurs less with vegetable oils because they have significantly lower levels of fatty acids than animal fats do. In addition, the higher temperatures associated with cooking with vegetable oils contribute to the faster breakdown of the soapy layer. Therefore, the extinguishing capability of the fire suppression system is reduced, and the probability of re-ignition is increased.

Locations of cooking greases, fats, and oils on shipboard settings

In general, cooking grease, fats, and oils should only exist in the galley and pantry areas onboard. Containers for cooking grease, fats, and oils, frying griddles, broilers, deep fat fryers, stoves, and ovens are major equipment containing or accumulating such liquids. Grease also accumulates in and around the range, particularly in the hoods, filters, the area immediately behind the filter screens (called the 'plenum'), and ductwork over galleys. The filters remove grease and oils from the smoke generated by the cooking process. When these substances remain on the filters, they react chemically. The reactions produce flammable substances, which are a fire hazard. In addition, clogged filters restrict the flow of ventilating air so that the temperature rises above the safe level. The result is usually a fire. When the range is being used for cooking, grease-laden air from the cooking area enters the ventilator duct. The air is forced to curve back and forth around several baffles at high speed. This zigzag motion tends to throw the grease and any lint and dust out of the air stream and onto the ducting and baffles. From there, it flows to a grease-collecting gutter. Fires in the cooking area can be serious. However, since they are out in the open, they usually can be extinguished completely. Fires in the plenum and the duct system are of most concern. Even after such fires are apparently extinguished, there may still be some fire hidden from view, or the fire may have extended out of the duct and into nearby compartments. For this reason, automatic fire extinguishing systems should be installed to protect all parts of the range, plenum, and ducts.

Extinguishing class 'F' or 'K' fires

To extinguish a fire created by auto-ignition of the cooking grease, fats, or oils, the flames must be extinguished, and the temperature of the burning liquid reduced below the auto-ignition temperature. The amount of heat involved with the liquid above 363 °C (685 °F) is high, and the use of the incorrect extinguisher can be extremely dangerous. For example, a water jet extinguisher directed at the surface of a burning cooking oil will create an explosion as the water is quickly converted into steam resulting in the expulsion of burning oil possibly spreading the fire and harming the operator. Conventional foam extinguishers have been proven to extinguish the flame, but the heat involved quickly destroys the foam blanket, exposing the surface of the oil, allowing re-ignition. Carbon dioxide and powder extinguishers are effective in extinguishing the flame, but without sealing the surface of the liquid from oxygen, the oil rapidly re-ignites. Conventional powder, foam, or CO_2 extinguishers are normally too powerful and direct and can easily splash the burning liquid and spread the fire. A fast high-rate discharge may be ideal for a petrol fire but is very dangerous for fires involving burning cooking oils or fats. Considering the aforementioned limitations of conventional powder, foam, or CO_2 extinguishers, or extinguishers designed for cooking oil, fires use wet chemical agents. Fire extinguishers of wet chemical type are available in hand portable models of 6 l (1.5 gal) and 9.5 l (2.5 gal). The extinguishing agent can be comprised of, but is not limited to, solutions of water and potassium acetate, potassium carbonate, potassium citrate, or a combination of these chemicals (which are conductors of electricity). The liquid agent typically has a pH of 9.0 or less. On Class 'F' or 'K' fires (cooking oil fires), the agent forms a foam blanket to prevent re-ignition. The water content of the agent aids in cooling and reducing the temperature of the hot oils and fats below their auto-ignition point. The agent, when discharged as a fine spray directly at cooking appliances, reduces the possibility of splashing hot grease and does not present a shock hazard to the operator. In recent years, the development of high-efficiency cooking equipment with high-energy input rates and the widespread use of vegetable oils with high auto-ignition temperatures have highlighted the need for a new Class 'F' or 'K' fire extinguisher. The wet chemical extinguisher was the first extinguisher to qualify to the new Class 'F' or 'K' requirements such as 'NFPA 10 – Standard for portable fire extinguishers', 'ANSI/UL711 – Rating and fire testing of fire extinguishers', and 'BS EN 3–7:2004+A1:2007 – Portable fire extinguishers – Part 7: Characteristics, performance requirements and test methods'.

In addition to offering rapid fire extinguishing effect, a thick foam blanket is formed to prevent re-ignition while cooling both the appliance and the hot cooking oil. Wet chemical extinguishers also offer improved visibility during firefighting as well as minimising cleanup afterward. For ship galleys with deep fat fryers, fire extinguishing systems and equipment for Class 'F' or 'K' fires are required in SOLAS regulation II-2/10.3 and 10.6.4.1 referring to IMO MSC.1/Circ.1275 'Unified Interpretation of SOLAS Chapter II-2 on The number and arrangement of portable fire extinguishers onboard ships' and 'ISO 15371:2009 – Ships and Marine technology – fire extinguishing systems for protection of galley cooking equipment'. Per IMO MSC.1/ Circ.1433 'For ships constructed before 1 July 2013, ISO 15371:2000, Fire extinguishing systems for protection of galley deep-fat cooking equipment – fire tests, may be used'. Moreover, exhaust ducts from galley ranges are to be fitted with fixed means for extinguishing a fire within the duct, when the exhaust duct passes through accommodation spaces or spaces containing combustible materials (refer to SOLAS regulation II-2/9.7.5 as amended by Resolution MSC.365(93)).

In this chapter, we have examined the five classifications of fires, the main sources of fuel for each class of fire, and the most common ways in which each class fire may start. In the next chapter, we will begin to turn our attention towards actual firefighting infrastructure commonly found onboard ships today.

Part 2

Fire extinguishing systems and infrastructure

Chapter 3

Shipboard fire main systems

In the first part of this book, we discussed the fundamental principles of combustion, which is commonly referred to as the science or chemistry of fire. It goes without saying that the science behind fire is much more complex and involved but what we have covered here is sufficient to provide a basic understanding of what fire is, what causes it, and what is needed to extinguish fires. In this part of the book, we turn our attention to the types of fire extinguishing systems and infrastructure commonly found on shipboard settings. To begin with, we will look at the shipboard fire main.

GENERAL PRINCIPLES OF THE SHIPBOARD FIRE MAIN SYSTEM

The shipboard fire main is a system consisting of sea inlet(s), suction piping, fire pumps, and a distributed piping system supplying fire hydrants, hoses, and nozzles located throughout the vessel. Its purpose is to provide a readily available source of water to any point throughout the vessel, which can be used to combat a fire and is considered the backbone of the firefighting systems onboard a vessel. Through the fire main system, the firefighter is provided with a reliable and versatile system capable of providing several different methods with which to engage a fire. Water can be supplied as a straight stream for combating deep-seated fires, as a spray for combating combustible liquid fires where cooling and minimum agitation is desired, or to protect personnel where cooling is the primary effect desired.

EXTINGUISHING CAPABILITIES OF WATER

Water primarily extinguishes a fire by the removal of heat (see Chapter 1). It absorbs heat more effectively than any other commonly used extinguishing agent due to its good thermal conductivity and its high latent heat of

DOI: 10.1201/9781003385523-5

vapourisation. It is most effective when it absorbs enough heat to raise its temperature to 100 °C (212 °F). At that temperature, water absorbs additional heat as it goes through the transition from a liquid to a vapour (i.e. steam). In the process of heating the water from normal temperatures, up through its conversion into steam, water absorbs approximately 2.6 kJ of heat per gram (1,117 BTU/lb) of water, which is a much higher heat absorption value than any other agent. This absorption of heat reduces the temperature of the burning vapours and reduces the amount of vapour being generated by the cooling of the fuel surface. With adequate cooling, there is insufficient heat to maintain the self-supporting combustion process and the fire goes out. Water also has an important secondary effect. When it turns to steam, it expands about 1,600 times in volume at atmospheric pressure. As a result, one cubic metre (cubic foot) of water can generate up to 1,600 cubic metres (cubic feet) of steam vapour. This great cloud of steam surrounds the fire, displacing the air that supplies oxygen for the combustion process. Thus, water provides a smothering action as well as cooling.

MOVING WATER TO THE FIRE

The purpose of the fire main system is simply to move the water to the locations needed at sufficient capacity and pressure so that it can be used as an effective extinguishing medium. At sea, the supply of water is limitless. However, moving the water is another matter. The amount of water that can be moved to a shipboard fire depends on the number and capacity of the fire pumps installed and the design of the fire main piping system. Even when water is available in vast quantities, it must still be used economically and wisely. If it is not, its weight can affect the equilibrium of the vessel. This is especially true if large amounts of water are introduced into, and remain at, high points in the vessel. The weight of the water raises the centre of gravity of the vessel and can impair the vessel's stability. Even water that is not confined, but can run to lower portions of the vessel, may affect the buoyancy of the vessel. Vessels have capsized and sank because excessive amounts of water were used during firefighting efforts. Each 1 m³ (35 ft³) of water weighs approximately one tonne (0.98 tonne). Onboard a vessel, water is transferred to the location of the fire in two ways:

- Via the fire main system, through hose lines that are manipulated by the vessel's personnel.
- Through piping systems that supply fixed manual or automatic sprinkler or spray systems.

Both are reliable methods for bringing water to bear on a fire, provided the pumps, piping, and all components of the system are properly designed and maintained. The first method of distribution, the fire main system, is discussed in this chapter. Water sprinkler and water spray systems will be discussed in Chapter 5. The fire main system is a vital part of the various firefighting systems installed onboard vessels. Since the hose and the nozzle complete the job of moving the water to the fire, it is equally important to understand the effects of the various operations of a hose line and the nozzles.

Straight streams

The straight stream, sometimes called the solid stream, is a valuable form of water for firefighting. The straight stream is formed by a nozzle that is specially designed for that purpose. The nozzle from which the water is thrown is tapered to increase both the velocity of the water at the discharge and the reach.

Efficiency of straight streams

The distance that a straight stream travels before breaking up or dropping is called its reach. Reach is important when it is difficult to approach close to a fire. Despite its name, a straight stream is not straight. Like any projectile, it has two forces acting upon it. The velocity imparted by the nozzle gives it reach, either horizontally or at an upward angle, depending on how the point man aims the nozzle. The other force, gravity, tends to pull the stream down where it encounters the deck. Accordingly, the stream is an arch. The maximum horizontal reach is then attained with the nozzle held at an upward angle of 35 to 40 degrees from the deck. Tables 3.1 and 3.2 provide information about the throw capability of individual nozzle sizes at various pressures/directions. Probably less than 10% of the water from a straight stream absorbs heat from the fire. This is because only a small portion of the water surface actually comes in contact with the fire, and only water that contacts the fire can absorb heat. The rest runs off, sometimes over the side, but more often the runoff becomes free-surface water and is a problem for the vessel.

Using straight streams

A straight stream should be directed into the seat of the fire. This is important since for maximum cooling, the water must contact the material that is burning. A solid stream that is aimed at the flames is ineffective. In fact, the main use of solid streams is to break up the burning material and penetrate to the seat of a Class 'A' fire. It is often difficult to hit the seat of a fire, even with the reach of a solid stream. Aboard a vessel, bulkheads with small openings can keep firefighters from getting into proper position to aim the stream into the fire. In some instances, there may be an obstruction between the fire and the point man. In these cases, frequently the stream can be bounced off a bulkhead or the overhead to get around the obstacle. This method can also be used to break a solid stream into a spray-type stream, which will absorb more heat. It is useful in cooling an extremely hot passageway that is keeping firefighters from advancing towards the fire.

The following are the basic characteristics of a straight stream:

- Can be aimed accurately.
- Has good reach.
- Must hit seat of fire to cool effectively.
- Run-off of water may be excessive.
- Generates very little steam.

Fog streams

The fog (or spray) nozzle breaks the water stream into small droplets. These droplets have a much larger total surface area than a solid stream. Thus, a given volume of water in fog form will absorb much more heat than the same volume of water in a straight stream due to the larger exposed surface area. The greater heat absorption of fog streams is important. Less water needs to be applied to remove the same amount of heat from a fire. In addition, more of the fog stream turns into steam when it hits the fire. Consequently, there is less runoff, less free-surface water, and less of a stability problem for the vessel. The cloud of steam surrounding the fire displaces the air that supplies oxygen for the combustion process. Thus, the fog stream helps to provide a smothering action as well as cooling. The following are the basic characteristics of a fog stream:

- Difficult to aim.
- Limited reach.
- Excellent cooling abilities.
- Generates steam.
- Has small amount of run-off.
- Pushes smoke and fire.
- Does not have to hit seat of fire to be effective.

There are essentially two different types of fog steams that are used onboard a vessel: the high velocity fog stream and the low velocity fog stream. Both can be used very effectively in combating a fire but serve different functions, as discussed later.

High-velocity fog streams

High-velocity fog streams are created by a specially designed hose nozzle. In addition to the direct cooling of the flame, the high-velocity fog stream can be used effectively to reduce heat in compartments, cabins, and cargo spaces.

High-velocity fog streams can also be used to move air in passageways and to drive heat and smoke away from advancing firefighters.

Low-velocity fog streams

Low-velocity fog is obtained by using a special applicator along with a combination nozzle. Applicators are tubes or pipes that are angled at 60 degrees or 90 degrees at the water outlet end. They are stowed for use with the low velocity head already in place on the pipe. Some heads are shaped somewhat like a pineapple, with tiny holes angled to cause minute streams to bounce off one another and create a mist. Some heads resemble a cage with a fluted arrow inside. The point of the arrow faces the opening in the applicator tubing. Water strikes the fluted arrow and then bounces in all directions, creating a fine mist. Low-velocity fog is effective in combating Class 'B' fires in spaces where entry is difficult or impossible. Applicators can be poked into areas that cannot be reached with other types of nozzles. They are also used to provide a heat shield for firefighters advancing with foam or high-velocity fog. Low-velocity fog can be used to extinguish small tank fires, especially where the mist from the applicator can cover the entire surface of the tank. However, other extinguishing agents, such as foam and carbon dioxide, are usually more effective.[1]

A water fog applicator might consist of a metal L-shaped pipe, the long limb being about 2 m (6.56 ft) in length, capable of being fitted to a fire hose and the short limb being about 250 mm (9.8 in) in length, fitted with a fixed water fog nozzle or capable of being fitted with a water spray nozzle.

Limitations of fog streams

Fog streams do not have the accuracy or reach of straight streams. While they can be effectively used on the surface of a deep-seated fire, they are not as effective as solid streams in breaking through and reaching the heart of the fire.

Combination nozzle operation

The combination or dual-purpose nozzle typically has a handle which is perpendicular to the plane of the nozzle and will produce either a straight stream or high-velocity fog, depending on the position of its handle. Combination nozzles are available for use with 3.8 cm (1.4 in) and 6.4 cm (2.5 in) hoses. Reducers can be used to attach a 3.8 cm (1.4 in) nozzle to a 6.4 cm (2.51 in) hose. A straight stream is obtained by pulling the handle all the way back to the operator. A fog stream is obtained by pulling the handle back halfway. The nozzle is shut down, from any open position by pushing the handle forward as far as it will go. In addition, combination nozzles typically have the means to attach a low-velocity applicator.

Straight stream and fog streams can be very effective against Class 'A' fires as they can penetrate the fire and cool the fuel surface. Although fog streams are generally better suited to extinguishing Class 'A' fires, they may also be used to good effect against Class 'B' fires.

Table 3.1 Water Discharge Rates

Rate (psi)	Gallons of water flowing per minute at various nozzle pressures and sizes												
	Nozzle sizes (inches)												
	$5/8$	$3/4$	$7/8$	I	$1^1/8$	$1^1/4$	$1^3/8$	$1^1/2$	$1^5/8$	$1^3/4$	$1^7/8$	2	$2^1/8$
30	63	91	124	162	205	253	307	353	429	498	572	651	826
32	65	94	129	167	212	261	317	377	443	514	591	673	854
34	67	97	133	172	218	269	327	389	457	530	610	693	880
36	69	100	137	177	224	277	336	400	470	546	627	713	905
38	71	103	140	182	231	285	345	411	483	561	645	733	930
40	73	106	144	187	237	292	354	422	496	575	661	752	954
42	75	108	147	192	243	299	363	432	508	589	678	770	978
44	77	111	151	196	248	306	372	442	520	603	694	788	1,000
46	79	113	154	200	254	313	380	452	531	617	710	806	1,021
48	80	116	158	205	259	320	388	462	543	630	725	824	1,043
50	82	118	161	209	265	326	396	472	554	643	740	841	1,065
52	84	121	164	213	270	333	404	481	565	656	754	857	1,087
54	85	123	167	217	275	339	412	490	576	668	769	873	1,108
56	87	125	170	221	280	345	419	499	586	680	782	889	1,129
58	88	127	173	225	285	351	426	508	596	692	796	905	1,149
60	90	130	176	229	290	357	434	517	607	704	810	920	1,168
62	91	132	179	233	295	363	441	525	617	716	823	936	1,187
64	93	134	182	237	299	369	448	533	627	727	836	951	1,206
66	94	136	185	240	304	375	455	542	636	748	850	965	1,224
68	96	138	188	244	308	381	462	550	646	750	862	980	1,242
70	97	140	190	247	313	386	469	558	655	761	875	994	1,260
72	98	142	193	251	318	391	475	566	665	771	887	1,008	1,278
74	100	144	196	2584	322	397	482	574	674	782	900	1,023	1,296
76	101	146	198	258	326	402	488	582	683	792	911	1,036	1,313
78	102	148	201	261	330	407	494	589	692	803	924	1,050	1,330
80	104	150	204	264	335	413	500	596	700	813	935	1,063	1,347
82	105	151	206	268	339	418	507	604	709	823	946	1,076	1,364
84	106	153	208	271	343	423	513	611	718	833	959	1,089	1,380
86	108	155	211	274	377	428	519	618	726	843	970	1,102	1,396
88	109	157	213	277	351	433	525	626	735	853	981	1,115	1,412
90	110	159	216	280	355	438	531	633	743	862	992	1,128	1,429

Theoretical discharge may be taken using this table.

For smaller and larger size nozzles (also straight bore type), use the following table:

N.P.	$1/4$	$5/16$	$3/8$	$1/2$	N.P.	$2^1/_2$	3
30	10	16	23	41	30	1,017	1,464
32	11	16	24	42	32	1,050	1,512
34	11	17	24	43	34	1,082	1,558
36	11	17	25	45	36	1,114	1,604
38	11	18	26	46	38	1,144	1,648
40	12	18	26	47	40	1,174	1,691
42	12	19	27	48	42	1,203	1,732
44	12	19	28	49	44	1,231	1,773
46	13	20	28	50	46	1,259	1,813
48	13	20	29	51	48	1,286	1,852
50	13	21	30	53	50	1,313	1,890
52	13	21	30	54	52	1,339	1,928
54	14	21	31	55	54	1,364	1,964
56	14	22	31	56	56	1,389	2,000
60	14	22	32	54	60	1,438	
70	16	24	35	62	70	1,553	
80	17	26	37	66	80	1,660	
90	18	28	40	70	90	1,761	
100	19	29	42	74	100	1,856	

For smooth bore nozzles 5/8" with 11/2" hose and 7/8" with 21/2" hose in still air conditions, the horizontal reach can be estimated at about 0.3 m (1 ft) per 0.07 bar (1 psi) with the Pitot pressure being between 14 and 15.9 m (47 and 52 ft) correspondingly at 3.5 bar (50 psi). For protection of Class 'B' fires (flammable liquids), where spray pattern is used, the range to be estimated is about 6–7.5 m (20–25 ft) in still air at 3.5 bar (50 psi) pressure. For smooth bore lines with 21/2" hose lines, Table 3.2 may be used for reference:

Table 3.2 Horizontal Reach of Water Streams

Rate (psi)	Hose streams throw (ft) for smooth bore lines with 21/2" hose lines											
	Nozzle sizes (inch)											
	$5/8$	$3/4$	$7/8$	1	$1 1/8$	$1 1/4$	$1 3/8$	$1 1/2$	$1 5/8$	$1 3/4$	$1 7/8$	2
30	18	25	35	45	57	70	85	101	118	138	158	180
32	19	27	37	48	61	75	91	108	126	147	168	192
34	20	29	39	51	65	80	97	115	134	156	180	204
36	21	31	41	54	68	85	102	121	142	165	190	216
38	22	32	44	57	72	90	108	128	150	174	200	228
40	23	34	46	60	76	94	113	135	158	184	210	240
42	25	36	48	63	80	99	119	142	166	192	221	252
44	26	37	50	66	84	103	124	148	174	202	232	264
46	27	39	53	69	88	108	130	155	182	211	242	276
48	28	41	55	72	91	112	136	162	190	220	253	288
50	29	42	58	75	95	117	141	168	198	229	263	300
52	31	44	60	78	99	122	147	175	206	239	274	312
54	32	45	62	81	102	126	153	182	214	248	285	324
56	33	47	64	84	106	131	158	189	221	256	295	336
58	34	49	66	87	110	136	164	196	229	266	305	348
60	35	51	69	90	114	141	170	202	237	275	316	360
62	36	53	71	93	118	145	176	209	245	284	326	372
64	38	54	73	96	122	150	181	216	253	293	337	384
66	39	56	76	99	125	155	187	223	261	303	347	396
68	40	58	78	102	129	160	193	230	270	314	359	409
70	41	59	80	105	133	164	199	237	277	322	370	421
72	42	61	83	108	137	169	204	244	285	331	380	433
74	44	63	85	111	141	174	210	250	294	340	390	445
76	45	65	87	114	144	179	216	257	301	350	402	456
78	46	66	90	117	148	183	222	264	309	360	412	469
80	47	68	92	120	152	188	227	270	317	358	422	480
82	48	70	94	123	156	193	233	277	325	377	432	492
84	49	71	96	126	159	197	238	284	333	386	444	505
86	50	73	99	129	163	202	244	291	340	395	454	518
88	52	75	101	132	167	207	250	297	348	405	465	529
90	53	76	103	135	171	212	255	304	357	415	475	540
92	54	78	106	138	175	216	262	311	365	423	485	552
94	55	80	108	141	179	220	267	318	372	432	495	565
96	56	82	110	144	182	226	272	325	380	441	506	578
98	57	83	112	147	186	230	278	332	389	451	517	590
100	59	85	115	150	190	235	284	338	397	460	530	600

In this chapter, we have examined the main infrastructure associated with the shipboard fire main system. In the next chapter, we will discuss the Class requirements for shipboard fire main systems.

NOTE

1 SOLAS (Consolidated Edition 2014) – footnote to II-2/10.5.5 concerning water fog applicators.

Chapter 4

Class requirements for fire main systems

In this chapter, we will discuss the Class requirements for shipboard fire main systems. The information provided in this chapter is general guidance only. Reference should always be made to the Class rules applicable to the specific vessel concerned for the complete set of current requirements.

MAIN FIRE PUMPS

Number required

Each vessel must be provided with at least two main fire pumps. For vessels 1,000 GRT and above, each main fire pump must be independently power-driven. The phrase 'independently power driven' should be understood to require that the driving power for each pump is not supplied from the vessel's propulsion unit and that the driving mechanism for each pump is independent of the other pump(s) driving mechanisms.

Arrangements

Unless an emergency fire pump is provided, the locations, as well as the arrangements, for access, power supplies, fuel supplies, sea suctions, suction and discharge piping, and supporting environmental services (lighting, ventilation, etc.) for the two main fire pumps must be arranged so that a fire in a space containing any one of the main fire pumps will not render both pumps inoperable. The following identifies the minimum requirements for access arrangements: separation and services supplying the two main fire pumps.

General arrangement and common boundaries

The pumps must be in separate rooms, completely isolated and independent from one another. It is usually necessary to confirm the arrangements to identify common entrances, passages, and structural isolation of common

DOI: 10.1201/9781003385523-6

boundaries. Only one common boundary is permitted between the spaces containing the two main fire pumps, and this single common boundary must be a class 'A-0' boundary or higher. Multiple common boundaries are not permitted.

Access

No direct access should be permitted between the spaces unless this is determined to be 'impracticable'. Where so determined to be 'impracticable', the Class rules will indicate where direct access may be considered. However, arrangements must comply with Class-mandated arrangements, which are outlined in the following:

- An arrangement where the access is by means of an airlock. One of the doors of the machinery spaces should be of A-60 class standard and the other should be of steel, reasonably gas-tight, self-closing, and without any hold-back arrangements.
- An access through a watertight door. This door should be capable of being operated from a space remote from the machinery space of category 'A' and the spaces containing the fire pumps. The remote control must also be located at a point that is unlikely to be cut off in the event of fire in those spaces, and for unattended propulsion machinery space operation, the door should be remotely operable from the firefighting station. As the door is a watertight door, controls from either side of the bulkhead are also required.
- Where direct access is provided in accordance with the arrangements specified in either of the two points above, a second means of access must be provided to the spaces containing the main fire pumps. The second means of access to each space must be totally independent of the space containing the other fire pump.

Power supplies

Power supplies for the two main fire pumps must be totally independent. If electrical, the review needs to confirm that the power is supplied from different sources (typically, one must be from the emergency generator), with the feeder cables being separated in all ways as far as practicable. Routing of the electric supply cable through the space containing the other fire pump or through high fire risk areas (Category 'A' machinery spaces) would not be acceptable. If diesel-driven, then arrangements for starting, fuel supply, air and exhaust piping, etc., are to be completely independent. Similarly, hydraulically driven pumps are to have independent power units and controls without interconnections.

Water supply

The sea chest, suction piping, and discharge piping arrangements to the fire main, as well as the pump and valve controls, must be arranged to be totally independent from the other pump and are not to be situated in or enter the space containing the other pump. The sea chest and suction pipe for one of the main fire pumps (or the emergency fire pump), which is located outside of the machinery space, may be situated in the machinery space provided the conditions and interpretations outlined by Class are satisfied. It should be noted that if the suction or discharge piping is permitted to be within a category 'A' machinery space, the piping should be all welded except for flange connections to the sea inlet valve.

Environmental systems

The arrangements of environmental systems (e.g., lighting and ventilation) should be completely independent for both pump locations. If the information available on hand is not sufficient to verify these arrangements, a comment should be made relative to the drawing and the installation verified by the attending Class Surveyor.

   (1)   Pump type, relief, etc.
       (a) Suitability.
            It is important to know certain details regarding the pumps (NPSH requirements, discharge characteristics, type of pump, etc.) being proposed for installation to verify the pump's suitability for fire main service. Part of determining the suitability of the pumps involves recognising that fire pumps must operate over a wide range of flow rates, from minimal a flow rate while supplying a single nozzle to full flow. With this wide range of operating flow rates, the resulting pressures developed by the pump must stay within an appropriate and safe range. Some pumps may be capable of providing the high flow rate at the minimum required pressure but may develop excessive pressures when delivering only the flow for a single nozzle. Such excessive pressures can make handling the hose very difficult and can present a danger if the pressure should be high and the hose gets loose. This issue should be addressed by Class, which indicates that the maximum pressure at any hydrant is not to exceed that at which effective control of a fire hose can be demonstrated. The pump characteristics play an important part in establishing compliance with this requirement. Other pumps may be able to provide suitable pressures at minimum flow rates but be unable to maintain the required pressures specified

by Class when delivering at the required full capacity flow rates. As a result of the need to supply a wide capacity range over a relatively narrow pressure band, centrifugal pumps with flat or very shallow head/capacity curves are usually proposed. Such pumps can supply minimal flow rates (i.e. approaching shut-off head) without creating excessive and dangerous pressures, yet are also capable of providing large flow rates while maintaining the required pressures at the most remote nozzles. Appropriate details verifying the suitability of the pumps should be requested and verified.

(b) Relief valves.
In addition to the concern associated with the possibility of danger in having excessive pressure for hose handling, many pumps are also capable of exceeding the maximum working pressures of the piping system components at shut-off pressures. Accordingly, Class requirements stipulate relief valves are provided in conjunction with all fire pumps if the pumps can develop a pressure exceeding the design pressure of the water service pipes, hydrants, and hoses. These relief valves are to be so placed and adjusted as to prevent excessive pressure in any part of the fire main system.

(c) Class certification.
In accordance with Class requirements, fire pumps are to be provided with Class-issued certificates. The certificates must be available for presentation to the attending Class Surveyor. Other firefighting service pumps, such as pumps for fixed water-based systems, or equivalent, local application firefighting systems, sprinkler systems, deck foam systems, etc., fall also in the Class certification requirements.

(d) Multiple service applications.
Pumps are not required to be dedicated for fire services only. They may be utilised for other purposes and designated accordingly (e.g. fire and ballast, fire and bilge, and general service pump). However, at least one pump must always be available. In addition, fire pumps should not be connected to systems normally used for transfer of oil. The Class rules do permit occasional, short-term oil duties for the pump. However, if so designated, there must be suitable positive changeover arrangements (e.g., elbow and blank flange) and a notice to assure the pump's connection to fire main after such use. Because of pumps' designation and use for other than fire services and the need for ready capability of the fire water supply, the pumps should have simple and reliable means to switch them to fire services.

The most common arrangement is two manifold valves at the pumps. In addition, there may be certain arrangements where the fire pumps also serve other firefighting services (e.g., sprinkler system), which must be immediately available. In such cases, isolation arrangements, such as swing elbows and blank flanges, would not be suitable.

(2)   Main fire pump capacities.

(a)   Total pump capacity.

In accordance with Class requirements, the combined output of the main fire pumps should be sufficient to deliver for firefighting purposes a quantity of water not less than four-thirds of the quantity required to be handled by each independent bilge pump as calculated in accordance with Class requirements. However, the total required capacity of the fire pumps need not exceed 180 m^3/h (792 gpm), insofar as classification requirements are concerned. To verify this requirement, the required independent bilge pump capacity must be established. Then the combined capacity of the fire pumps, while delivering at the pressure necessary to meet the requirements set by Class at the most remote hydrants (i.e. pressure drops associated with flowing through the main must be considered), must not be less than four-thirds of that value.

(b)   Individual pump capacity.

In accordance with Class requirements, each of the main fire pumps is required to have a capacity of not less than the greater of:

- 80% of the total required capacity divided by the number of required fire pumps.
- 25 m^3/h (110 gpm).
- The capacity required to deliver at least the two jets of water.

Class requirements often stipulate that these fire pumps are to be capable of supplying the fire main system under the required conditions, which should be understood to be those pressures specified by Class. Most Class rules identify two interrelated requirements, that being the *flow rate* at a *required pressure*. Therefore, when evaluating the individual fire pump capacity, the review should establish that each fire pump can deliver the largest of the above required flow rates through the most hydraulically remote hydrants while maintaining the pressures specified by Class at the nozzles. In the case of the third requirement, the largest size nozzles to be used should be taken into consideration. In a few instances, more than two pumps might be designated for fire services. The Class rules leave this situation to special consideration. For example, provision of two smaller capacity pumps arranged together to make up the capacity of one of the required pumps will satisfy the intent of the Class rules, provided they are arranged with common suction/discharge valves, power supply, start/stop controls, etc., and are interconnected in such a way as to perform the functions of the required pump. In accordance with Class rules, each pump for fire extinguishing, which is installed in addition to the required number of pumps, should have a capacity of at least 25 m3/h (110 gpm) and should be capable of delivering at least the two jets of water required by Class. Per IACS UI SC270, onboard cargo ships designed to carry five or more tiers of containers on or above the weather deck, the total capacity of the main fire pumps need not exceed 180 m3/h in cases where the mobile water monitors, required by SOLAS regulation II-2/10.7.3.2. are supplied by separate pumps and piping system.

(3) Ready availability of water supply.

It should be noted that Class rules require that cargo vessels be provided with the means to provide the ready availability of water supply for spaces intended for centralised or unattended operation.

(4) Connections to fire main from another engine room pump.

The Class rules also require that in addition to the main and emergency fire pumps, on a cargo vessel where other pumps (such as general service, and bilge and ballast) are fitted in a machinery space, arrangements are to be made so that at least one of these pumps, having the capacity and pressure required by Class, is capable of providing water to the fire main.

EMERGENCY FIRE PUMP

If the arrangements for the two main fire pumps are such that a fire in one space can put both main fire pumps out of service, an emergency fire pump is required to be provided, as per the Class rules. The emergency fire pump is required to be a fixed independently driven pump and the requirements for the pump and its location are identified by Class. These requirements are summarised below.

Arrangement (independent systems)

As per Class, the emergency fire pump system, including the power source, fuel oil supply, and electric cables, as well as the lighting and ventilation for the emergency fire pump space are to be independent of the main fire pumps so that a fire in any one compartment will not render both the main and emergency fire pumps inoperable. These items are to be verified during plan review whenever possible. If the fire pump is electrically driven, the power should be supplied from the emergency generator and the feeder cable routed outside of high fire risk areas (category 'A' machinery spaces). In addition, the electrical power and control cables for the emergency fire pump and valves are to also be widely separated from any electrical and control cables associated with the main fire pumps.

Capacity

In association with the requirements specified by the relevant Class rules, the capacity of the pump is not to be less than 40% of the total capacity of the fire pumps required, and in any case, not less than the following:

- For cargo vessels of 2,000 GRT and upward: 25 m³/h (110 gpm).
- For cargo vessels less than 2,000 GRT: 15 m³/h (66 gpm).

Where applicable, the emergency fire pump is also to be capable of simultaneously supplying the amount of water needed for any fixed fire extinguishing system protecting the space containing the main fire pump. The emergency fire pump must also be suitable for the service and be provided with adequate protection against over pressurisation. Per IACS UI SC270, onboard cargo ships designed to carry five or more tiers of containers on or above the weather deck, the total capacity of the emergency fire pump need not exceed 72 m3/h.

Pressure

When the pump is delivering the quantity of water discussed earlier, the pressure at any hydrant should be not less than the minimum pressures given in the relevant Class rules. Verification of compliance with this requirement must be based upon considering the worst-case situation.

Starting arrangements

Any diesel-driven power source for the pump should be capable of being readily started in its cold condition down to a temperature of 0 °C (32 °F) by hand (manual) cranking. If this is impracticable, or if lower temperatures are likely to be encountered, heating arrangements are to be provided so that ready starting will be assured. If hand (manual) starting is impracticable, other means of starting may be considered. These means are to be such as to enable the diesel-driven power source to be started at least six times within a period of 30 minutes and at least twice within the first ten minutes. Diesel engines exceeding 15 kW (20 hp) are to be equipped with an approved auxiliary starting device (e.g., starting battery, independent hydraulic starting system, or independent starting air system). Also, any internal combustion engine driving an emergency fire pump should be capable of being readily started in its cold condition down to a temperature of 0 °C (32 °F). If this is impracticable, or if lower temperatures are likely to be encountered, heating arrangements are to be provided for ready starting of generator sets.

Fuel supply

Any service fuel tank must contain sufficient fuel to enable the pump to run on full load for at least three hours, and sufficient reserves of fuel are to be available outside the main machinery space to enable the pump to be run on full load for an additional 15 hours.

Suction

The total suction head and net positive suction head of the pump are to be such that the Class requirements are obtained under all conditions of list, trim, roll, and pitch likely to be encountered in service. This excludes the docking condition of the vessel. In selecting the emergency fire pump, the minimum available net positive suction head should provide a safety margin of at least the 1 m (3.3 ft) or 30% of required net positive suction head of the pump, whichever is less. In addition, the sea valve should be operable from a position near the pump.

Arrangement (boundaries)

The space containing the emergency fire pump is not to be contiguous to the boundaries of Category 'A' machinery spaces or those spaces containing main fire pumps. Where this is not practicable, the common bulkhead between the two spaces should be insulated to a standard of structural fire protection equivalent to that required for a control station. The common bulkhead must be constructed to 'A-60' class standard and the insulation should extend at least 450 mm (18 in) outside the area of the joint bulkheads and decks. The structural fire protection arrangements of the compartment containing the emergency fire pump should also be verified.

Supply and delivery piping

The emergency fire pump, including the seawater suction and delivery pipes, are be located outside the compartment containing the main fire pumps. The sea chest with valve and main part of the suction piping should be, in general, outside the machinery spaces containing the main fire pumps. If this arrangement is impractical, the sea chest may be fitted in the machinery spaces containing the main fire pump on the condition that the suction valve is remotely controlled from a position near the pump in the same compartment and that the suction pipe is as short as possible. Only short lengths of suction and discharge piping may penetrate the machinery spaces if enclosed in a steel casing.

Access

Like the requirements discussed earlier for the access arrangements between the main fire pumps, no direct access is permitted between the machinery space of Category 'A' and the space containing the emergency fire pump and its source of power. This requirement assumes that the main fire pumps will be in a Category 'A' machinery space. Accordingly, should the main fire pumps not be in a Category 'A' machinery space, it should be noted that direct access between the space containing the emergency fire pump and the main fire pump(s) also would be unacceptable. However, like the arrangements for main fire pumps, where this is unavoidable, the Class rules do permit the following arrangements to be considered:

- *Via airlock*. An arrangement where access is provided by means of an airlock may be considered. For such an arrangement, one of the doors of the machinery spaces should be of A-60 class standard and the other should be of steel, reasonably gas-tight, self-closing, and without any hold back arrangements.
- *Via watertight door*. An access through a watertight door may also be considered. This door should be capable of being operated from a space remote from the machinery space of Category 'A' and the space containing the emergency fire pump, and should be unlikely to cut off in the event of fire in those spaces.
- *Via second means of access*. Where direct access is provided in accordance with the arrangements specified in point 1 or point 2, a second means of access must be provided to the spaces containing the emergency fire pumps. The second means of access must be totally independent of the space containing the main fire pumps.
  Related Class interpretation/instruction. Where an emergency fire pump room (EFPR) is accessible from the steering-gear room (SGR), and access to the SGR is from a machinery space of category 'A' or a

space containing the main fire pumps, the arrangement will be acceptable based on the following conditions:

- A Class technical office must be satisfied with the shipyard's or designer's technical explanation that the EFPR arrangement described earlier is unavoidable (i.e. impracticable for the emergency fire pump room to be located elsewhere).
- The requirements of Class are to be satisfied.
- A-60 insulation with 450 mm (17.7 in) overlap must be provided to contiguous boundaries between EFPR and machinery spaces of Category 'A' or those spaces containing main fire pumps.
- Where the second protected means of access required is provided from EFPR to directly outside of the SGR (i.e. via an escape trunk), then A-0 insulation is acceptable between SGR and machinery spaces of Category 'A' and those spaces containing main fire pumps. However, the protected escape trunk must be at least A-60. Where the second means of escape is provided from the EFPR to SGR and then from SGR to the outside of SGR, the above A-0 insulation should be replaced by A-60 insulation.
- EFPR ventilation may be common with SGR, provided the SGR ventilation is separate and independent from the ventilation for the machinery spaces of category 'A' or those spaces containing main fire pumps.
Related IACS Unified Interpretation.
UI SC 114 re. SOLAS (Consolidated Edition 2014) Chapter II-2, Regulation 10.2.2.3.2.1.
When a single access to the EFPR is through another space adjoining a machinery space of Category 'A' or the spaces containing the main fire pumps, class A-60 boundary is required between that other space and the machinery space of category 'A' or the spaces containing the main fire pumps.

(1) *Ventilation.* Ventilation arrangements to the space containing the independent source of power for the emergency fire pump are to be such as to preclude, as far as practicable, the possibility of smoke from a machinery space fire entering or being drawn into that space. The space should be well ventilated, and power for mechanical ventilation should be supplied from the emergency source of power. This should be verified by the Class plan review, or the attending Class Surveyor requested to verify the compliance.

(2) *Priming.* Class rules often require that the emergency fire pump should be of the self-priming type and the same should be confirmed at the plan review.

(3) *Class certification.* The requirement for Class certification of fire pumps, as discussed in Chapter 3, is also applicable to emergency fire pumps.

FIRE MAIN SIZING AND SYSTEM PRESSURES

The ability of the fire main system to provide adequate quantities of water through a sufficient number of fire hoses at the necessary pressures is critical to the effectiveness and usefulness of system. Accordingly, the ABS Class rules stipulate (as an example) the following:

4–7–3/1.7.1 Fire main diameter

The diameter of the fire main and water service pipes should be sufficient for the effective distribution of the maximum required discharge from two fire pumps operating simultaneously. However, the diameter need only be sufficient for the discharge of 140 m3/h (616 gpm).

4–7–3/1.7.2 Fire main pressure

With the two pumps simultaneously delivering through nozzles specified in 4–7–3/1.15, the quantity of water specified in 4–7–3/1.7.1, through any adjacent hydrants, the following minimum pressures are to be maintained at all hydrants:

- *Vessels of 6,000 gross tonnage and upwards: 0.27 N/mm2 (2. kgf/cm2, 40 psi)*

- *Vessels less than 6,000 gross tonnage: 0.25 N/mm2 (2.6 kgf/cm2, 37 psi)*

In examining the requirements stipulated in the Class rules, there are several interrelated issues involved with these requirements, including:

- Total required flow rate for the fire main system.
- Pressure drops associated with the fire main distribution system.
- Specific hydrants being considered.
- Sizes of the hose nozzles used.
- Performance characteristics of the pumps.

Each of these issues impacts the ability of the system to perform as required, and the interrelationship of these items needs to be well understood when evaluating the system.

(1) *Required quantity of water discharged.* The first item is the total required amount of water that the system must be capable of discharging from the fire main. A certain amount of clarification is needed. First, the requirements set out by the ABS in SVR 4–7–3/1.7.2 reference the quantity of water identified in SVR 4–7–3/1.7.1. Next, the quantity

of water identified in SVR 4–7–3/1.7.1 is the 'maximum required discharge from two fire pumps operating simultaneously'. Defining the term 'maximum required discharge from two fire pumps operating simultaneously' should be considered as the greater of the following:

- The total required capacity of the combined fire main pumps calculated in accordance with SVR 4–7–3/1.3.1; or
- The combined capacities of the individual fire pumps where each individual fire pump capacity is determined in accordance with SVR 4–7–3/1.3.2.

However, SVR 4–7–3/1.7.1 does place a limit on the capacity required by indicating that the diameter of the fire main and distribution system need only be sufficient for the discharge of 140 m³/h (616 gpm), similar to the provisions of SOLAS. Accordingly, the required amount of water that the system must be capable of continuously discharging is the lesser of the 'maximum required discharge from two fire pumps operating simultaneously' or 140 m³/h (616 gpm). However, it should be noted that there are certain Flag administrations that do not recognise the upper limit specified in SOLAS.

(2) *Fire main diameter/pressure drop.* In association with other items, the pressure drop within the fire main when discharging the quantity of water discussed earlier is dependent upon the diameter, length, etc. of the fire main distribution piping, size and length of the hoses, nozzle sizes, and elevation of hydrants. The requirements for the sizing of the fire main distribution piping are provided in SVR 4–7–3/1.7.1, which specifies 'the diameter of the fire main and water service pipes are to be sufficient for the effective distribution of the maximum required discharge from two fire pumps operating simultaneously'. Accordingly, the fire main piping must be sized to effectively deliver the 'maximum required discharge from two fire pumps operating simultaneously'. As noted earlier, SVR 4–7–3/1.7.1 does place a limit on the capacity requirement by indicating that the diameter of the fire main and distribution system need only be sufficient for a maximum discharge of 140 m³/h (616 gpm).

(3) *Hydrants involved.* SVR 4–7–3/1.7.2 indicates that the pressures stated therein must be maintained at 'all' hydrants when discharging the total quantity of water through adjacent hydrants. To confirm that the minimum pressure will be available at 'all' hydrants, the review must consider the worst-case scenario. This would be at that hydrant which is the most hydraulically remote from the pumps. The 'most hydraulically remote' hydrant is that hydrant in the system which would result in the largest pressure drop for the system when discharging the total required flow rate through the adjacent nozzles. In considering this requirement, one needs to recognise that the most hydraulically remote hydrant may

be the hydrant located the farthest distance from the pump. However, it could also be the hydrant located highest on the superstructure or it may be a hydrant located at the end of a poorly sized branch line. It is a function of the pressure drop (at the required flow rate) in the piping to that hydrant which of course is a function of the distance, height, and sizing of the fire main system leading to that hydrant. During the review of the fire main system, one should verify that with the total required quantity of water flowing through the nozzles of hoses attached to adjacent hydrants, the minimum specified pressure is maintained at all hydrants, including the most hydraulically remote hydrant.

(4) *Hose nozzles.* As indicated in SVR 4–7–3/1.7.2, the quantity of water involved must be discharged through hose nozzles that are sized in accordance with SVR 4–7–3/1.15. This is important to note since the size of the nozzles to be used will impact the pressure drop through the system. A larger nozzle orifice will have a smaller pressure drop than a small nozzle when trying to flow the same amount of water. This requirement is also important in establishing that the discharge of the water through the adjacent hydrants must be through hoses which are attached to the hydrants being used to discharge the required flow rate. Thus, any calculations must to take into account resistance of fire hoses with attached nozzles.

(5) *Pump performance.* The heart of the system is clearly the fire pump. Accordingly, the performance characteristics of the fire pumps at the required flow rates will be one of the major factors in the ability of the fire main system to provide the required minimum pressures. At least basic details of the type of pump and copies of the pump curves are necessary to verify compliance with SVR 4–7–3/1.7.1 and 4–7–3/1.7.2.

(6) *Conclusion.* Typically, calculations are required to verify compliance with SVR 4–7–3/1.7.1 and 4–7–3/1.7.2. The calculations should establish that the fire main system, which includes the pumps, fire main distribution piping, hoses, and nozzles, are properly selected and sized to supply the required total flow rate to the most hydraulically remote hydrants while maintaining the required pressure at every hydrant. The review of the calculations should confirm that the designer truly understands the requirements and allows the engineer to identify any problem areas before the system is built. However, even if verified by calculations, the ability of the system to meet the requirements of SVR 4–7–3/1.7.2 should also be demonstrated to the satisfaction of the attending Class Surveyor.

FIRE HOSE REACTION

The Class rules stipulate that the maximum pressure at any hydrant is not to exceed that at which the effective control of a fire hose can be demonstrated. Accordingly, the maximum pressure at any hydrant should be limited to a pressure for which effective control of the fire hose can be maintained. The following provides a method for estimating the reaction forces from a discharging hose. Typically, forces from the nozzle with pitot tube pressures up to 6.9 bar (100 psi) would not be considered to produce excessive reactions. Acceleration of water discharged from the smooth bore nozzle creates a reaction force:

$$F = 1.57D2 \ plbf$$

$$F = 0.157D2 \ pN$$

where
F = reaction force, N (lb)
D = orifice diameter, mm (in)
p = pilot pressure, bar (psig)

The maximum pressure at any hydrant and the associated reaction forces should be investigated during the review of the calculations under all operating scenarios. This may require additional calculations since during the calculations discussed in this chapter are based upon full-flow calculations at the most remote hydrants. However, the maximum hydrant discharge pressure will typically exist with a single nozzle discharging at the hydrant hydraulically closest to the pumps.

ISOLATION VALVES AND ROUTING ARRANGEMENTS

In accordance with Class rules, isolating valves are to be fitted in the fire main risers leading from those fire pumps located within category 'A' machinery spaces, as well as in any fire main branch piping, which is routed back into category 'A' machinery spaces. Because of the uncertain conditions or intensity of fire in such spaces, the presumption is that the space is inaccessible, equipment inoperable, and the main is left open-ended (e.g., damage and maintenance). As a result, water supply from the fire pump or pumps will not be able to build up the required pressure at hydrants and, further, such conditions will result in uncontrolled flooding of the space unless accessible isolation arrangements are available. Therefore, the Class rules require accessible (i.e. located outside of the category 'A' machinery space) valves be fitted so that damaged or open sections of the fire main piping located within the category 'A' machinery spaces can be isolated to eliminate the pressure loss in the system. In considering the fire main system design, particular attention should be paid to the way the fire main branches are routed. Class requires that the fire main and branches are to be so arranged that when the isolating valves are shut, all the hydrants on the vessel, except those into any specific category 'A' machinery space, can be supplied with water by a fire pump not located in this machinery space and through pipes which do not enter the isolated space. This allows isolation of the piping within the space and use of the hydrants nearby to fight the fire. Therefore, the isolation valves must be located:

- Where the fire main risers leave the category 'A' machinery spaces containing the fire pumps.
- Upstream of the point where any fire main branch piping re-enters the space.

As a result, no branch downstream of the riser piping isolation valve(s) should re-enter the space without an isolation valve, and no branch from the isolated space should project outside or into an adjacent compartment without an isolation valve and second source of supply.

FIRE MAIN PIPING COMPONENTS AND MATERIALS

All valves, fittings, and piping are to comply with the applicable requirements set by Class. Accordingly, all valves and fittings are to be designed and constructed in accordance with a recognised standard, be suitable for the intended pressures, and comply with all other requirements mandated by the Class rules. A submitted drawing should include a complete Bill of Materials that provides the material specifications, standards of construction, pressure ratings, and types of all valves and fittings, as well as the material specification, sizes, and wall thickness of the piping. The drawing should also identify the maximum system pressure and provide adequate information to identify the layout and arrangement of the system(s). In addition to the common requirements and limitations set by Class, the materials used in the system are not to be rendered ineffective by heat, unless the components are adequately protected. This requirement is important so that the fire main system will remain intact and functional even if one portion of the fire main piping is within the immediate vicinity of the fire. To be considered not 'readily rendered ineffective by heat', a component should be certified as having passed an acceptable recognised fire test or the material should have a melting temperature higher than the test temperature specified in an acceptable fire test. Insofar as classification requirements are concerned, the fire test requirement may be waived, if shipyards/manufacturers/designers (as appropriate) document that all components used in fire mains have a solidus melting temperature above the minimum temperature specified in an acceptable fire test. The requirement for the fire main material not to be rendered ineffective by heat extends from the pumps throughout the system to the fire station hydrant valves. The hoses and nozzles are not required to meet the above criteria as it is assumed that the integrity of the system will be maintained by the hydrant valve, but are required to be of approved non-perishable material, as discussed later. The integrity of the valves to maintain closure is also critical to avoid losses of fire main pressure. For this reason, resiliently seated valves may be considered for use in fire main systems, provided they can pass an appropriate fire test acceptable to Class requirements and can be effectively closed with the resilient seat damaged or destroyed such that leakage through the closed valve is insignificant. No hose connections, non-metallic expansion joints, and similar provisions are allowed. Frequently used rubber (cloth inserted) gaskets are also not permitted, as their destruction under the fire could lead to the loss of pressure in the system. For regulatory compliance, criteria stipulated by the Flag State administration must be satisfied and may differ in criteria from that stated earlier.

HYDRANT LOCATIONS AND FIRE HOSES AND NOZZLES

Hydrant locations

The Class rules require that

> the number and position of the fire hydrants be such that at least two jets of water not emanating from the same hydrant (one of which should be from a single length of hose) may reach any part of the vessel normally accessible to the passengers or crew while the vessel is being navigated and any part of any cargo space when empty, any RORO cargo space or any special category space (in which latter case the two jets are to reach any part of such space, each from a single length of hose). Furthermore, such hydrants are to be positioned near the accesses to the protected spaces.

Class also requires that at least two combination solid stream and water spray nozzles and hoses sufficient in length to reach any part of the helicopter deck be provided at any helicopter facilities, which would include landing as well as winching facilities. In addition to the spacing to facilitate hose access, the Class rules specify that the piping and hydrants are to be so placed that the fire hoses may be easily coupled to them. Also, where deck cargo is carried, hydrants are to be located such as to avoid damage during cargo operations, but at the same time to always remain easily accessible. Verification that the arrangements of the hydrants comply with Class requirements during plan review can be difficult to achieve as the actual structural arrangements, location of equipment, etc., are less reliable, and therefore, is typically included in the inspection undertaken by the Class

Surveyor. However, the layout and arrangement of the fire stations, as provided in the drawing, in conjunction with the proposed hose lengths should be reviewed and any areas of concern identified. For those areas which are questionable or where the details provided on the drawings are insufficient to verify compliance, appropriate comments should be made to the issue to the Class Surveyor's attention.

Related IMO MSC/Circ.1120 Interpretation

SOLAS (Consolidated Edition 2014) Chapter II-2/10.2.1.5.1: 'Location of hydrant in machinery spaces'.

At least one hydrant with hose, nozzle, and coupling wrench should be provided in machinery spaces of Category 'A'.

(1) Hydrant valves.

In accordance with SVR 4–7–3/1.11.2, a valve is required to be fitted at a fire station to serve each fire hose so that any fire hose may be removed while the fire pumps are at work. Normally, each hydrant is fitted with an individual valve, but the drawing should be reviewed to verify such compliance and the hose valve determined to be of a design and material which will not be rendered ineffective by heat.

(2) Fire hoses.

(a) Hose approval.

Fire hoses are required by the FSS Code to be of approved non-perishable material. Documentation to verify the material as non-perishable must be via certified to recognised standard by competent independent testing laboratory. In addition, fire hoses are to be Class Type Approved or Type Approved by the vessel's Flag State administration.

(b) Hose length.

Hoses are to be sufficient in length to project a jet of water to any of the spaces in which they may be used, but they are to have a length of at least 10 m (33 ft). Fire hoses should not have a length greater than:

- 15 m (49 ft) in machinery spaces.
- 20 m (66 ft) in other spaces and open decks.
- 25 m (82 ft) on open decks with a maximum breadth in excess of 30 m (98 ft).

The limitation on the length of hose is due to difficulties in unfolding, kinking, and handling of hoses.

(c) Number of hoses.

The Class rules do not specifically require a hose to be provided at each fire hydrant location. The minimum number of hoses to be provided on vessels of 1,000 GRT and upwards is at least one for each 30 m (100 ft) length of the vessel and one spare, but in no case less than five in all. Also, this number does not include any hoses required in any engine or boiler room. The number of hoses in the machinery space(s) is not specified but must comply with the relevant government authorities. However, while the Class rules do not require a hose to be provided at each fire hydrant location, they do indicate that unless one hose and nozzle are provided for each hydrant on the vessel, there should be complete interchangeability of the hose couplings and nozzles. Vessels carrying dangerous goods should be provided with three additional hoses and three additional nozzles.

(3) Nozzles.

The Class rules stipulate that the standard nozzle sizes are to be 12 mm (0.5 in), 16 mm (0.625 in), and 19 mm (0.75 in) or as near thereto as possible. Larger diameter nozzles may be permitted, provided the system can maintain the required pressures. For nozzles for accommodation and service spaces, the Class rules usually indicate that a nozzle size greater than 12 mm (0.5 in) need not be used. For machinery spaces and exterior locations, the nozzle sizes are to be such as to obtain the maximum discharge possible from two jets at the pressure mentioned in the Class rules from the smallest pump, provided that a nozzle size greater than 19 mm (0.75 in) need not be used. All nozzles are to be of a dual-purpose type (i.e. combination spray/jet type) incorporating a shut-off. In addition, all nozzles are to be of an approved type (which implies approval by a competent authority) for the intended service. Also, nozzles made of plastic material such as polycarbonate may be accepted subject to review of their capability and serviceability as marine use fire hose nozzles.

Related IACS Unified Interpretation.

UI SC146 – SOLAS (Consolidated Edition 2014) Reg. II-2/10.2.3.

Aluminium alloys may be used for fire hose couplings and nozzles, except in open deck areas of oil tankers and chemical tankers.

(4) Miscellaneous.

The Class rules also require each hose to be provided with a nozzle and the necessary couplings, and that the fire hoses together with any necessary fittings and tools are to be kept ready for use in conspicuous positions near the water service hydrants or connections. Details of the fire stations normally provide sufficient details to verify compliance with this requirement.

INTERNATIONAL SHORE CONNECTION ARRANGEMENTS

Each vessel is required to be provided with at least one international shore connection and arrangements are to be provided to enable the connection to be used on each side of the vessel. The purpose of the connection should provide a uniform piping joint, which allows water supply to the vessel's fire main from another vessel or shore facility. For this purpose, at least one hydrant on either side of the vessel, suitable for such connection, should be designated an international shore connection station. Provision of dedicated port/starboard hydrants with shore connections permanently fixed to them is an acceptable arrangement. The Class rules address the dimensions and design of the international shore connection. The standard dimensions of flanges for the international shore connection are required to be in accordance with the Class rules for that vessel. For guidance, a copy of the rules set by the ABS is provided in Table 4.1. In addition, the Class rules indicate that the connection should be of steel or other suitable material and should be designed for 10 bar (150 psi) services. The flange should have a flat face on one side and on the other side, it should be permanently attached to a coupling that will fit the vessel's hydrant and hose. The connection should be kept aboard the vessel together with a gasket of any material suitable for 10 bar (150 psi) services, together with four 16 mm (0.62 in) bolts, 50 mm (1.9 in) in length, and eight washers. The connection should be accompanied with a gasket, bolts, nuts, and washers. The shore connection should be positioned in a readily accessible location.

Table 4.1 Dimensions of International Shore Connections

	SI and MKS units	**US equivalent units**
Outside diameter	178 mm	7 in
Inside diameter	64 mm	2.5 in
Bolt circle diameter	132 mm	5.2 in
Slots in flange	Four holes – 19 mm in diameter – spaced equidistantly on a bolt circle of the above diameter slotted to the flange periphery.	Four holes – 0.75 in in diameter – spaced equidistantly on a bolt circle of the above diameter slotted to the flange periphery.
Flange thickness	14.5 mm minimum	0.57 in minimum
Bolts and nuts	Four each of 16 mm diameter, 50 mm in length	Four each of 0.63 in diameter, 1.97 in in length

COLD WEATHER PROTECTION

The arrangements of the fire main piping and hydrants are to be such as to avoid the possibility of freezing. Accordingly, those sections of the fire main, which may be subject to freezing temperatures, are required to be provided with adequate protection. Examples of acceptable arrangements would be heat tracing, installation of isolation valves inside the superstructure, and provision of drain valves (for non-pressurised system).

ADDITIONAL REQUIREMENTS FOR VESSELS WITH AUTOMATION NOTATIONS

In vessels with a periodically unattended machinery space or when only one person is required on watch, the Class rules requires that there must be immediate water delivery from the fire main system at a suitable pressure. This can be accomplished by either the remote starting of one of the main fire pumps from the navigation bridge and fire control station, if any, or permanent pressurisation of the fire main system by one of the main fire pumps. This requirement may be waived for cargo vessels of less than 1,600 GRT, provided the arrangement of the machinery space access makes it unnecessary. Where the option of remote starting is selected, the remote operation of any, and all, valves necessary to allow the remotely started fire pump to supply the fire main should be included (e.g. open those valves to the fire main and to close those to the other systems, and sea water supply valves). Furthermore, the Class rules typically require provisions to be made for remotely starting one of the main fire pumps on the navigation bridge, unless the fire main is permanently pressurised. It also requires means to be provided for the starting of one of the main fire pumps at the fire control station, if fitted. As an alternative, means may be provided at firefighting station to start the emergency fire pump, to satisfy this requirement. For vessels seeking the ABS 'ACCU' notation, a firefighting station is required to be provided. Starting one of the fire pumps located outside the propulsion machinery space, including operation of all necessary valves, to pressurise the fire main from the firefighting station is required. However, starting of one of the main fire pumps is still required to be provided on the navigation bridge. When the fire main is permanently pressurised by one of the fire pumps and the pump is arranged to start automatically upon the loss of pressure in the system (caused by opening a hydrant), no additional pump starting arrangements would be required.

ALTERNATIVE REQUIREMENTS FOR STEEL VESSELS UNDER 90 M IN LENGTH AND LESS THAN 1,000 GRT

The following identifies alternatives to certain requirements discussed earlier.

(1) Fire pumps.

The number of pumps required is still the same, that is, two pumps. However, only one of the pumps is required to be independently operable.

(2) Fire main pressure.

For vessels of less than 1,000 GRT, the minimum fire main pressure should be sufficient to produce two 12 m (40 ft) jet throws through any adjacent hydrants to any part of the vessel.

(3) Hoses.

In cargo vessels of less than 1,000 GRT, the number of fire hoses should be calculated in accordance with the provisions of SOLAS regulation II-2/10.2.3.2.3.1. However, the number of hoses should in no case be less than three.

ADDITIONAL REQUIREMENTS FOR OIL AND FUEL OIL TANKERS

The following requirements are in addition to those discussed earlier. On oil and fuel oil tankers, it is usual for Class to require additional isolation valves to be fitted in the fire main at the poop front in a protected position and on the tank deck at intervals not more than 40 m (131 ft) to isolate damaged sections of the fire main. These isolation valves should be located immediately forward of any hydrant position which will allow the ability to isolate a damaged portion of the main forward of a hydrant and still be able to use the closest intact hydrants.
Related IACS Unified Interpretation.
UI SC146 – SOLAS regulation II-2/10.2.3.
Aluminium alloys may be used for fire hose couplings and nozzles, except in open deck areas of oil tankers and chemical tankers.
Related IMO Unified Interpretation.
MSC/Circ.1456 as amended by MSC.1/Circ.1492 to SOLAS regulation II-2/10.2.1.4.4 'Location of the fire main isolation valves in tankers'.

The complete interpretation of the phrase 'the isolation valves shall be fitted in the fire main at the poop front in a protected position' would be that the valve should be located within an accommodation space, service spaces, or control station. However, the valve may be located on the open deck aft of the cargo area provided that the valve is located:

- At least 5 m (16.4 ft) aft of the aft end of the aftermost cargo tank; or
- If the above is not practical, within 5 m (16.4 ft) aft of the aft end of the aftermost cargo tank provided the valve is protected by a permanent steel obstruction.

ADDITIONAL REQUIREMENTS FOR PASSENGER VESSELS

Passenger vessels present a unique risk insofar as fire hazards are concerned. Unlike the situation of a cargo vessel where the crew are the only personnel onboard and are typically well-trained, a passenger vessel will be carrying many untrained, and sometimes less mobile, people than the personnel normally found onboard. Accordingly, the risk of loss of life is great, and therefore, passenger vessels are required to meet special, more demanding fire main system requirements. The fire pumps, fire mains, hydrants, and hoses for passenger vessels are to be in accordance with requirements specified by Class. The following requirements are mandatory:

(1)   Fire pumps.
  (a)   Number of fire pumps.
        Vessels of 4,000 GRT and over are to be provided with at least three independently driven fire pumps and vessels of less than 4,000 GRT are to be provided with at least two independently driven fire pumps.
  (b)   Capacity of fire pumps.
        The fire pumps are to be capable of delivering for firefighting purposes a quantity of water not less than two-thirds that required to be dealt with by the bilge pumps when employed for bilge pumping at the appropriate pressure. This should be verified by confirming that the combined discharge capacity of the proposed fire pumps is at least two-thirds of the combined capacities of the required bilge pump (three or four bilge pumps, depending on the bilge pump numeral).
  (c)   Arrangement of fire pumps and power sources.
        Sea connections, fire pumps, and their source of power are to be located aft of the collision bulkhead and are to be such that:

- In vessels of 1,000 GRT and upwards, a fire in any one compartment will not put all the fire pumps out of action.
- In vessels of less than 1,000 GRT, if a fire in any one compartment could put all pumps out of action, an emergency fire pump complying with Class minimum requirements should be provided.

(d) Arrangement of fire water supply.

The arrangements for the ready availability of water supply are to comply with the following:

- In vessels of 1,000 GRT and upwards, at least one effective jet of water should be immediately available from any hydrant in an interior location and so to provide the continuation of the output of water by the automatic starting of a required fire pump.
- In vessels of less than 1,000 GRT, an effective stream should be readily available, either by the automatic starting of at least one fire pump or by remote starting of at least one fire pump from the navigation bridge.
- If fitted with periodically unattended machinery spaces, provisions for fixed water fire extinguishing arrangement for such spaces equivalent to those required for normally attended machinery spaces are to be provided.

(2) Fire mains.

(a) Size.

The diameter of the fire main and water service pipes should be sufficient for the effective distribution of the maximum required discharge from two fire pumps operating simultaneously. If the fire pump capacities differ, the largest two should be selected in determining compliance.

(b) Pressure.

With the two pumps simultaneously delivering through nozzles, the quantity of water through any adjacent hydrants, the following minimum pressures are to be maintained at all hydrants:

- 4,000 GRT and upwards 0.4 N/mm^2 (4.1 kgf/cm^2, 58 psi)
- Less than 4,000 GRT 0.3 N/mm^2 (3.1 kgf/cm^2, 44 psi).

However, the maximum pressure at any hydrant is not to exceed that at which the effective control of a fire hose can be demonstrated.

Related IMO MSC Circ.1120 Interpretation.

SOLAS regulation II-2/10.2.1.2.1.1, 'Automatic starting of fire pumps and prevention of freezing in pipes'.

Special attention should be given to the design of the continuously pressurised pipelines for prevention of freezing in pipes in vessels entering areas where low temperatures may exist.

(3)   Hydrants.

The number and position of hydrants are to be such that at least two jets of water not emanating from the same hydrant (one of which should be from a single length of hose) may reach any part of the vessel normally accessible to the passengers or crew while the vessel is being navigated, and any part of any cargo space when empty, any RORO cargo space or any special category space (in which latter case the two jets will reach any part of such space, each from a single length of hose). Furthermore, such hydrants are to be positioned near the accesses to the protected spaces. For installations in the accommodation, service and machinery spaces, the number and position of hydrants are to be such that the Class requirements are complied with when all watertight doors and all doors in main vertical zone bulkheads are closed. Where access is provided to a machinery space of Category 'A' at a low level from an adjacent shaft tunnel, two hydrants are to be provided external to, but near the entrance to, that machinery space. Where such access is provided from other spaces, in one of those spaces, two hydrants are to be provided near the entrance to the machinery space of category 'A'. Such provision need not be made where the tunnel or adjacent spaces are not part of the escape route.

(4)   Fire hoses.

At interior locations in vessels carrying more than 36 passengers, the fire hoses are to be connected to the hydrants at all times. There should be at least one fire hose for each hydrant, and these hoses are to be used only for the purposes of extinguishing fires or testing the fire extinguishing apparatus at fire drills and surveys. Vessels are to be provided with fire hoses; the number of and diameter of which are to be to the satisfaction of Class.

(5)   Nozzles.

In vessels carrying more than 36 passengers, each machinery space of Category 'A' should be provided with at least two suitable water fog applicators.

(6)   Location/arrangement of water pumps for other fire extinguishing systems.

Pumps required for the provision of water for other required fire extinguishing systems, their sources of power, and their controls are required to be installed outside the space or spaces protected by such systems and are to be so arranged that a fire in the space or spaces protected will not put any such system out of action.

ADDITIONAL/ALTERNATIVE REQUIREMENTS FOR RORO VESSELS

Due to the hazards associated with carrying vehicles with fuel remaining in their tanks, as well as the potential for additional amounts of combustibles in the vehicle construction and their cargo, the Class rules provide for certain requirements which must be complied with:

(1) Locations of hydrants.
    The number and position of the fire hydrants on any vessel are to be such that at least two jets of water not emanating from the same hydrant (one of which should be from a single length of hose) may reach any part of the vessel. Normally for RORO vessels, the Class rules also require that two jets of water, each from a single length of hose, are to be capable of reaching any part of RORO cargo space. Plan review should verify that the hydrant locations within the RORO spaces comply with this requirement.

(2) Low velocity water fog applicators.
    Each RORO cargo space intended for the carriage of motor vehicles with fuel in their tanks for their own propulsion should be provided with at least three water fog applicators. The plan review of the fire main system should confirm that the appropriate number of water fog applicators is provided for each RORO space.

ADDITIONAL/ALTERNATIVE REQUIREMENTS FOR GAS CARRIERS

Application

All vessels, irrespective of size, carrying products, which are subject to IMO gas code requirements, must comply with the requirements of SOLAS regulations II-2/10.2, 10.4, and 10.5. Accordingly, the review of the fire main system should verify compliance with the SOLAS regulations, as adopted by the Class rules.

Fire pump/fire main sizing

When the fire pump and fire main are used as part of the water spray system as permitted by Class, the required fire pump capacity and fire main and water service pipe diameter should not be limited by the provisions of the SOLAS regulations II-2/10.2.2.4.1 and II-2/10.2.1.3. Accordingly, where the fire pump and fire main are used as part of the water spray system pressure, the calculations should be based upon both systems operating simultaneously, each at their required maximum discharge capacity.

Pressure

The Class rules specify that the discharge pressure requirements of SOLAS regulations II-2/10.2.1.6 should be met at a pressure of at least 5.0 bar (72 psi) gauge. The discharge pressures specified By Class should be identical to those found in SOLAS regulation II-2/10.2.1.6. Accordingly, the values indicated by Class should be replaced with a pressure of at least 5.0 bar (72 psi) gauge and the system pressure calculations reviewed accordingly.

Arrangements

As required by Class, the arrangements are to be such that at least two jets of water can reach any part of the deck in the cargo area and those portions of the cargo containment system and tank covers above the deck. Further, the locations of the fire hydrants are to satisfy the above requirement, as well as the requirements in SOLAS regulations II-2/10.2.1.5.1 and II-2/10.2.3.3, with a hose length as specified in SOLAS regulation II-2/10.2.3.1.1. The review should seek to verify the same, but frequently the need for Class Surveyor verification of this item is needed.

Isolation valves

Stop valves are required to be fitted in any crossover line for the fire main. In addition, stop valves are also to be fitted in the fire main or mains at the poop front and at intervals of not more than 40 m (131.2 ft) between hydrants on the deck in the cargo area for the purpose of isolating damaged sections of the main.

Nozzles

All water nozzles provided for firefighting use are required by Class to be of an approved dual-purpose type capable of producing either a spray or a jet.

Piping materials

All pipes, valves, nozzles, and other fittings in the firefighting systems are required to be resistant to corrosion by seawater (for which purpose galvanised pipe, e.g., may be used) and to the effect of fire.

Remote operations

Where the vessel's engine room is unattended, the Class rules require that arrangements be made to start and connect at least one fire pump to the fire main by remote control from the navigation bridge or other control station outside the cargo area. Details regarding this arrangement should be verified during plan review or a statement made requesting the Class Surveyor to verify the same.

STEEL VESSELS UNDER 90 M (295 FT) IN LENGTH

Certain provisions are made for vessels less than 90 m (295 ft) in length in consideration of their unique services and arrangements. The requirements applicable to steel vessels under 90 m in length and being classed for unrestricted service are discussed below.

(1) *Vessels >500 GRT and unrestricted service.* The fire main system of vessels under 90 m (295 ft) and over 500 GRT which are to be classed for unrestricted service are to comply with the requirements identified in the Class rules pertaining to vessels <90 m (295 ft). In general, the requirements for the fire main systems of vessels under 90 m (295 ft) are very similar to the requirements discussed earlier for vessels greater than 90 m (295 ft). Therefore, the discussion provided in Chapter 3 of this book for vessels greater than 90 m (295 ft) is applicable, along with certain clarifications or modifications as provided below.

(2) *Vessels <500 GRT.* The fire main system of vessels under 90 m and less than 500 GRT which are to be classed for unrestricted service are to comply with the requirements identified in the Class rules pertaining to vessels <90 m (295 ft). In general, these requirements are the same as the requirements outlined earlier for vessels 500 GRT or more with the following exceptions:

   (a) *Fire pumps.* Power-driven fire pumps must be capable of producing a 12 m (40 ft) jet throw through any two adjacent hydrants and only one of the pumps needs to be independently power driven. For vessels less than 20 m (65 ft), one power-driven pump, which may be an attached pump, is required along with one hand pump. The hand pump should have a capacity of at least 1.1 m3/h (5 gpm). Table 4.2 lists the capacity of the power-driven pump:

   (b) *Fire main.* The fire main should be sized for the effective distribution of the maximum required discharge of the pump(s) as per the Class rules for vessels <90 m (295 ft) in length.

   (c) *Hoses and nozzles.* Hoses and nozzles are to comply with the above-noted requirements, except that one fire hose with couplings and nozzles should be provided for each hydrant and one spare hose should be carried onboard. Further, hose diameters need not exceed 38 mm (1.5 in). In addition, hoses for vessels

Table 4.2 Pump Capacities

Length of vessel	Minimum pump capacity
Less than 20 m or 65 ft	5.5 m³/h (25 gpm)
20 m (65 ft) or greater, but less than 30.5 m (100 ft)	11.0 m³/h (50 gpm)
30.5 m (100 ft) or greater, but less than 61 m (200 ft)	14.3 m³/h (66.6 gpm)
Greater than 61 m (200 ft)	Refer to the Class rules for vessels under 90 m (295 ft) in length

less than 20 m (65 ft) may be of good commercial grade having a diameter of not less than 16 mm (0.625 in) and are to have a minimum test pressure of 10.3 bar (150 psi) and a minimum burst pressure of 31.0 bar (450 psi).

HIGH-SPEED CRAFT

The fire main system requirements for high-speed craft can be found in the Class rules for each specific vessel. For vessels classed by ABS, these are in Part 4, Chapter 5 of the *ABS Rules for Building and Classing High-Speed Craft* (HSCR). Cargo craft which proceed more than eight hours from a place of refuge are required to comply with separate Class rules. However, high speed craft which do not proceed, in the course of their voyage for more than 8 hours, at operational speed, and from a place of refuge, can usually confirm their compliance with the Class rules by complying with alternative requirements. Accordingly, this distinction must be established prior to conducting the review of the fire main system on high-speed craft.

(1) Materials.
As indicated in Class rules, materials readily rendered ineffective by heat are not permitted to be used for fire mains unless adequately protected. To be considered not 'readily rendered ineffective by heat', a component must be certified as having passed an applicable recognised fire test, or the material should have a melting temperature higher than the test temperature specified in an applicable fire test. For more information, refer to Chapter 3.

(2) Fire pumps.
(a) Number of pumps.
In accordance with Class rules, all high-speed craft are to have at least two fire pumps. For craft of 500 GRT and above, the

pumps are to be independently power-driven. For craft less than 500 GRT, only one of the pumps needs to be independently power-driven and one of the pumps may be attached to the propulsion unit. For craft less than 24 m (79 ft) in length, one power-driven pump, which may be an attached unit, and one hand-operated fire pump are permitted.

(b) Type of pumps.

Like the requirements for other vessels, the Class rules will indicate which sanitary, ballast, bilge, or general service pumps may be accepted as fire pumps, provided that they are not normally used for pumping oil. If the pumps are subject to occasional duty for the transfer or pumping of fuel oil, changeover arrangements that prevent operation for firefighting when configured for fuel transfer are to be fitted.

(c) Pump capacity.

• High-speed craft of 500 GRT and above.

Each of the power-driven fire pumps required by the Class rules should have a capacity of not less than two-thirds of the quantity required to be dealt with by each of the independent bilge pumps, but not less than 25 m3/h (110 gpm), and in any event should be capable of delivering at least the two required jets of water. These pumps are to be capable of supplying the water under the required conditions. Where more pumps than required are installed, their capacity will be subject to special consideration.

• Craft less than 500 gross tonnes.

The capacity of each power-driven fire pump should be in accordance with the above or other Class rule(s), whichever is less. Hand pumps, where permitted, should have a minimum capacity of 1.1 m3/h (5 gpm).

(d) Pressure.

The Class rules require all power-driven fire pumps to have sufficient pressure to simultaneously operate the adjacent hydrants.

(e) Relief valves.

In conjunction with all fire pumps, relief valves are to be provided if the pumps can develop a pressure exceeding the design pressure of the water service pipes, hydrants, and hoses. These valves are to be placed and adjusted to prevent excessive pressure in any part of the fire main system. In general, the relief valve should be set to relieve at no greater than 1.7 bar (25 psi) in excess of the pump pressure necessary to maintain requirements.

(3)   Arrangement.
    (a)   Standard arrangement.
        For craft of 500 GRT and above, the two main fire pumps (including their power source, fuel supply, electric cable and lighting, and ventilation for the spaces in which they are located) are to be in separate compartments so that a fire in any one compartment will not render both main pumps inoperable. Only one common boundary is allowed between the compartments in which case the single common boundary should be at least to A-0 standard. No direct access is allowed between the compartments, except that where this is impracticable, an access meeting the requirements in Chapter 3 may be considered.
    (b)   Alternative arrangement.
        Where it is impracticable to do otherwise, a direct access between the compartments containing the main fire pumps may be considered provided:
        •   A watertight door capable of being operated locally from both sides of the bulkhead and from a safe and accessible location outside of these spaces is provided. The means for the latter operation is expected to be available in the event of fire in these spaces; or
        •   An air lock consisting of two gas-tight steel doors. The doors are to be self-closing without any hold back arrangements.
        •   In addition to the arrangements specified in points 1 and 2, a second protected means of access should be provided to the space containing the fire pumps.

(4)   Isolation.
    For craft of 500 GRT and above, isolating valves and other arrangements, as necessary, are to be provided so that if a fire pump and its associated piping within its compartment are rendered inoperable, the fire main can be pressurised with a fire pump located in another compartment.

(5)   Fire main.
    (a)   Size.
        Class rules require the diameter of the fire main and water service pipes to be sufficient for the effective distribution of the maximum required discharge from two fire pumps operating simultaneously, notwithstanding that the diameter must be sufficient for the discharge of 140 m³/h (616 gpm). Refer to Chapter 3 for related discussion concerning the sizing of the fire main and the required calculations.

(b) Valves.

Class rules require that a valve be fitted to serve each fire hose so that any fire hose may be removed while the fire pumps are at work. These would be the hydrant valves.

(c) Cold weather protection.

Clearly, portions of the fire main which will be exposed to environmental conditions can be subject to freezing, and therefore, Class rules require fire main systems to be provided with drains, circulation loops, or other means for cold weather protection.

(6) Hydrants.

(a) Number and position.

Class rules will dictate whether the number and position of the hydrants are to be such that at least two jets of water not emanating from the same hydrant (one of which should be from a single length of hose) may reach any part of the craft. A review of the drawings by the engineer may be able to verify compliance but typically verification by the Class Surveyor is required.

(b) Materials.

Like the requirement stated in Chapter 3, materials of the hydrants are not to be readily rendered ineffective by heat.

(c) Installation.

The pipes and hydrants are to be placed so that the fire hoses may be easily coupled to them. In craft where deck cargo may be carried, the positions of the hydrants are to be such that they are always readily accessible, and the pipes are to be arranged to avoid risk of damage by such cargo.

(7) Hoses.

The Class rules require that the fire hoses are to be of a type certified by a competent independent testing laboratory as being constructed of non-perishable material to a recognised standard. The hoses are to be sufficient in length to project a jet of water to any of the spaces in which they may be required to be used. The maximum length of hose is not to exceed 23 m (75 ft). Class also requires each hose to have a nozzle and the necessary couplings. Fire hoses, together with any necessary fittings and tools, are to be kept ready for use in conspicuous positions near the hydrants.

(a) Diameter.

The Class rules indicate that the hoses are not to have a diameter greater than 38 mm (1.5 in). However, for craft less than 20 m (65 ft) in length, a hose of good commercial grade having a diameter of not less than 16 mm (0.62 in), a minimum test pressure of 10 bar (150 psi), and a minimum burst pressure of 31 bar (450 psi) would be permitted.

(b) Number.
The Class rules require that one fire hose with the couplings and nozzle should be provided for each hydrant. Additionally, at least one spare hose should be kept onboard. Compliance with these items should be verified during the plan review.

(8) Nozzles.
In accordance with the Class rules, the standard nozzle sizes are to be 12 mm (0.5 in), 16 mm (0.625 in), and 19 mm (0.75 in), or as near thereto as possible. Larger diameter nozzles may be permitted, subject to compliance with the specific Class rules for that vessel and equipment. For accommodation and service spaces, a nozzle size greater than 12 mm (0.5 in) is not required. For machinery spaces and exterior locations, the nozzle size should be such as to obtain the maximum discharge possible from the two jets at the pressure specified for the smallest pump. However, a nozzle size greater than 19 mm (0.75 in) is not required. All nozzles are to be of an approved dual-purpose type (i.e. spray and jet type) incorporating a shut-off. Fire hose nozzles of plastic type material such as polycarbonate may be accepted, subject to review of their capacity and serviceability as marine fire hose nozzles.

CARGO VESSELS FOR RIVER SERVICE

Fire main systems installed on vessels intended to operate only on rivers and intracoastal waterways, and receiving the classification notation for the restricted service, are required to comply with Class rules, for example, the *ABS Rules for Building and Classing Steel Vessel for Service on Rivers and Intracoastal Waterways*.
Accordingly:

(1) Fire pumps.
As indicated in the Class rules for cargo vessels on river service, all self-propelled vessels are to be fitted with at least one fire pump. For vessels over 20 m (65 ft), the pump should be power-driven. Vessels under 20 m (65 ft) must carry a hand operated pump, where the fire pump is allowed to be hand-operated, it must have a capacity of at least 1.1 m³/h (5 gpm) and should be equipped with suitable suction and discharge hoses for use on firefighting. Sanitary, ballast, bilge, and general service pumps may be accepted as fire pumps. However, each pump is required to be capable of providing a full supply of water to the fire hoses whereby at least two powerful streams' throw of 12 m (40 ft) can be rapidly and simultaneously directed into any part of the vessel.

(2)   Relief valves.
Relief valves are required to be fitted on power-driven fire pumps in accordance with Class rules, unless it can be shown that the arrangements are such as to prevent excessive pressure in any part of the fire main system.

(3)   Fire main.
The vessel is required by Class to be fitted with a fixed fire main system, and it should be sized for the simultaneous operation of at least the two fire hoses.

(4)   Locations of hydrants.
The number and position of the fire hydrants are to be such that at least two streams of water not emanating from the same hydrant may reach any part of the vessel. One of the streams should be from a single length of hose not more than 23 m (75 ft) long for 38 mm (1.49 in) hose or 15 m (50 ft) long for 63 mm (2.48 in) hose. In addition, piping and hydrants are to be placed so that fire hoses may be easily coupled to them, and where deck cargo is carried, hydrants are to be located such as to avoid damage during cargo operations, but at the same time to remain easily accessible. Verification that the arrangements of the hydrants comply with the above requirements during plan review can be very difficult as the actual structural arrangements, location of equipment, etc. make this approach much less reliable. Therefore, the arrangements of the hydrants are to be verified by the Class Surveyor. However, the layout and arrangement of the fire stations as provided in the drawing, in conjunction with the proposed hose lengths, should be reviewed and any areas of concern identified. For those areas which are questionable or where the details provided on the drawings are insufficient to verify compliance, appropriate comments should be made to issue to the Class Surveyor's attention.

(5)   Materials.
The materials used in the system are not to be rendered ineffective by heat unless the components are adequately protected. This requirement is important so that the fire main system will remain intact and functional even if one portion of the fire main piping is within the immediate vicinity of the fire. For more information about this issue, refer to Chapter 3.

(6)   Hydrant valves.
A valve is required to be fitted to serve each fire hose so that any fire hose may be removed while the fire pumps are at work. Normally, each hydrant is fitted with an individual valve, but the drawing should be reviewed to verify such compliance and the hose valve determined to be of a design and material which will not be rendered ineffective by heat.

(7) Fire hoses.
Fire hoses are to be of approved material. In addition, the minimum hose diameter for all vessels over 21 m (65 ft) in length should be not less than 38 mm (1.5 in) diameter. For vessels 20 m (65 ft) and under, 19 mm (0.75 in) diameter hose may be used. The hoses are to be sufficient in length to project a jet of water to any of the spaces in which they may be required, but the maximum length of hose is not to exceed 23 m (75 ft). The Class rules do not require a hose to be provided at each fire hydrant location. The minimum number of hoses to be provided is at least one for each 30 m (100 ft) length of the vessel and one spare. Also, this number does not include any hoses required in any engine or boiler room. The number of hoses in the machinery space(s) is not specified, suggesting this should comply with the relevant governmental authorities. Each hose should be provided with a nozzle and necessary couplings. Unless one hose and nozzle are provided for each hydrant on the vessel, there should be complete interchangeability of the hose couplings and nozzles.

(8) Nozzles.
The minimum internal diameter of any hose nozzle is not to be less than 16 mm (0.62 in), except as discussed below. Nozzles for hoses attached to hydrants serving machinery spaces are required to be suitable for spraying water on oil, or alternatively, they are to be of the dual-purpose type. For vessels of 100 GRT and less, the minimum internal diameter of the nozzles may be 8 mm (0.31 in). For vessels of less than 20 m (65 ft), garden-type hoses may be used.

ADDITIONAL REQUIREMENTS FOR PASSENGER VESSELS IN RIVER SERVICE

In addition to the requirements discussed in Chapter 3, river vessels involved in passenger service must comply with the following.

(1) Fire pumps.
Two independently power-driven fire pumps are required for passenger vessels, one of which should be dedicated to firefighting duties at all times. Arrangement of the pumps, sea suctions, and sources of power should be such that a fire or causality in any one space would not render both pumps inoperable. Where the arrangements are such that a fire would put both main fire pumps out of service, an emergency fire pump should be fitted. Where an emergency fire pump is fitted, it should be a fixed independently driven pump with

a capacity of not less than 25 m³/h (110 gpm) and be capable of simultaneously delivering 12 m (40 ft) jet throw from any two adjacent hydrants. The arrangements for the sea connections are to be outside the machinery space. The required pumps are to be capable of delivering at a pressure of at least 3.0 bar (44 psi) a quantity of not less than two-thirds the quantity required to be dealt with by the bilge pumps.

(2) Location of hydrants.

The number and position of hydrants should be such that at least two streams of water, not emanating from the same hydrant may be directed at any part of the vessel normally accessible to the passengers and crew, any part of the cargo space, any RORO space, and any special category space. Each stream should be from a single length of hose. Furthermore, hydrants are to be located near the accesses to protected spaces. In accommodation, service and machinery spaces, the number and position of hydrants are to be such that the above requirement can be complied with when all watertight doors and all doors in main vertical bulkheads are closed.

(3) Fire hoses.

For passenger vessels carrying more than 36 passengers, fire hoses are to be connected to the hydrants at all times in interior locations.

(4) Nozzles.

Standard nozzle sizes are to be 12 mm (0.5 in), 16 mm (0.625 in), and 19 mm (0.75 in), or as near to as possible. For accommodation and service spaces, a nozzle size greater than 12 mm (0.5 in) are not required. For machinery spaces and exterior locations, the nozzle size should be such as to obtain the maximum discharge possible from two jets at the referenced pressures from the smallest pump. However, a nozzle size greater than 19 mm (0.75 in) is not required.

YACHTS

The fire main system requirements for yachts are found in Part 4, Chapter 5 of the *Guide for Building and Classing Yachts (Yacht)*.

(1) Fire pumps.

Yachts are required to carry two fire pumps. However, only one of these pumps need be independently power-driven, and therefore, one of the fire pumps may be attached to the propulsion unit. Where approved by Class, yachts may use sanitary, bilge, and general service pumps as fire pumps.

(2)   Fire pump pressure.
Power-driven fire pumps are to produce sufficient pressure to supply the effective stream required by Class from any hydrant on the
yacht. An effective stream is normally considered to be a 12 m (40
ft) throw through the nozzle and hose provided.

(3)   Fire main.
The above fire pumps should normally supply the water through a
fixed fire main system.

(4)   Hoses, nozzles, and hydrants.
In accordance with the Class rules specific to yachts, hoses are not to
have a diameter greater than 38 mm (1 in). Hoses for vessels less than
20 m (65 ft) in length may be of a domestic service type of good commercial grade having a diameter of not less than 16 mm (0.625 in).

Nozzle sizes are to be in accordance with Class requirements and
as discussed in Chapter 3 of this book. The nozzles attached to
hydrants serving machinery spaces of yachts over 20 m (65 ft) are
to be of the type suitable for spraying water on oil or dual-purpose
nozzles. The fire hydrants are to be of sufficient number and so
located that any part of the yacht may be reached with an effective
stream of water from a single length of hose not exceeding 15 m (50
ft). The review of the fire hydrant locations and hoses should try to
establish the same, and comment made in the review letter requesting Class Surveyor verification was necessary. Whilst not generally
specified in most Class rules pertaining to yacht classification, the
materials of the fire main system should not be rendered ineffective
by heat. In addition, isolation valves are not specifically required
on the fire main. However, installation of isolation valves is highly
desirable in controlling the fire water supply.

FISHING VESSELS

The requirements for the classification of fishing vessels are specifically outlined in the Class rules for fishing vessels, or if not, in the Class rules for
vessels of L<90 m (295 ft). Where the former applies, the Class rules do not
usually provide any specific requirements for the fire main system on fishing vessels. Despite this, the attention of owners, designers, and builders is
directed to the regulations of government and other authorities dealing with
active fire protection for fishing vessels. Where authorised by the administration of a country signatory to the *International Conference on Safety of
Fishing Vessels, 1977/1993 Protocol*, and upon request of the owners of an
existing vessel or a vessel under construction, Class will usually review plans
and survey the vessel for compliance with the provisions of the 1977/1993

Protocol and certify thereto in the manner prescribed in the Protocol. In such cases, the IMO *Torremolinos International Convention for the Safety of Fishing Vessels, 1977*, as modified by the *Torremolinos Protocol of 1993* should be consulted as follows:

- Chapter V – Fire protection, fire detection, fire extinction and firefighting.
- Part A – General
- Part B – Fire safety measures in vessels of 60 m (197 ft) in length and over.
- Part C – Fire safety measures in vessels of 45 m (148 ft) in length and over but less than 60 m (197 ft) in length.

ACCOMMODATION BARGES

One effective jet of water must be immediately available from any hydrant in an interior location and to provide the continuation of the output of water by the automated starting of a required fire pump. Two independently power-driven fire pumps are required, each arranged to draw directly from the sea and discharge into a fixed fire main. However, where there are high suction lifts, reference should be made to the Class rules pertaining to the design and construction of accommodation barges. One of the required pumps should be dedicated to firefighting duties at all times. Arrangement of the pumps, sea suctions, and sources of power are to be such that a fire or casualty in any one space would not render both pumps inoperable. Where shore supply of water is available for firefighting purposes, the requirements for fire pumps should comply with the relevant Class rules.

Related Class Interpretation/Instruction (I/I).

ABS Guide for Building and Classing Accommodation Barges.

Note 1: For accommodation barges that are to be supplied with shore-side water for firefighting purposes, it is required that two sources of water be provided. As such, in applying A1/3.1 of the Accommodation Barge Guide, an independently driven emergency fire pump of the capacity specified in this Chapter 4 are to be provided regardless of the supplied shore side pressure.

Note 2: An emergency fire pump, as required by A1/3.1 of the Accommodation Barge Guide, can be replaced by dedicated shore side fire station facilities provided the alternative arrangements listed below are complied with:

- The fire station facilities are dedicated to the Naval Facility that the barge is located in, they are no more than 1.5 mi (2.5 km) from the barge, and the response time of the fire station (the time between the activation of the alarm and the arrival of the fire engine at the location of the barge) should be no more than five minutes.

- The fire engine should have a pump capacity and the equipment necessary to meet the Accommodation Barge Guide requirements of 3/5.3.1 at the most extreme locations of the barge.
- The fire engine should be capable of using sea water as an alternative water supply to the shore hydrant connection.
- The fire engine should be capable of connecting directly to the barge's fire main.
- The fire hydrants on the quay are to be spaced no more than 15.25 m (50 ft) apart and are to have a dedicated water supply.
- While the barge is moored to a quay, it will be connected to the quay fire main, which will be capable of continuously providing 6.2 bar (90 psi) water to the barge fire main.
- Should the quay fire main water pressure drop below 6.2 bar (90 psi) while the barge is manned, an alarm should sound in the control station and the dedicated shore-based fire station should be notified and put on stand-by status.
- The shore side fire hydrants are to be fitted with bleeder valves to maintain the flow of water in freezing conditions.
- A non-return valve should be fitted to prevent water pumped from the fire engine into the barge fire main from flowing back into the shore water supply piping.
- A clear description of the secondary firefighting arrangements should be posted in the control station and included on the fire safety plan.
- The subject barges are to be Classed 'Accommodation Barge – Restricted Service' and the Record entry should include note 26, which states 'Certain systems and arrangements accepted at the request of the US Government'.

Note 3: For an accommodation barge or a hotel barge tied up at a quay/riverbank/land mass or attached to a series of other barges, a 'dry' fire main system without an automatically starting fire pump is acceptable as a special modified system for limited or restricted service under the provisos of 1/5.3 and A1/1 of the Accommodation Barge Guide subject to compliance with the following list:

- The barge is classed for river service and so distinguished in the Record.
- The firefighting system complies with the published requirements of the governmental authority in which the vessel is registered.
- Pipe, pump, and valve arrangements are to ensure an immediate supply of water at the required pressure from any hydrant in an interior

location by suitably placed remote starting arrangements of a required fire pump.

- A ready means of escape (e.g. gangway or equivalent) from the barge is provided for safe egress of personnel.
- A public address system or other effective means of communication is available throughout the accommodation, service spaces, and control stations.
- An efficient patrol system is maintained so that an outbreak of fire may be promptly detected.
- Fixed fire detection and fire alarm systems are provided as per 4.13 of the Accommodation Barge Guide.

Pump requirements

The required pumps must meet the following:

- Total capacity must comply with 3/5.3.1 of the Accommodation Barge Guide.
- Each pump's capacity must comply with 3/5.3.1 of the Accommodation Barge Guide.
- Remote start-up of the required pumps and remote operation of the suction and discharge valves are necessary where either pump is located in a space not normally manned and is relatively far from working areas, as per 3/5.3.1 of the Accommodation Barge Guide.
- Other pumps may be accepted as fire pumps, provided that they are not normally used for pumping oil and one pump is always dedicated to firefighting duties.
- Non-return valves must be fitted on every centrifugal pump, which is connected to the fire main.
- Relief valves are needed as per 3/5.3.1 of the Accommodation Barge Guide.

Fire main and pressure

A fixed fire main should be provided and be so equipped and arranged as to meet the following requirements and the requirements of 3/5.3.1 of the Accommodation Barge Guide:

- The fire main must be of sufficient diameter to comply with 3/5.3.1 of the Accommodation Barge Guide.
- Adequate pressure must be maintained as per 3/5.3.1 of the Accommodation Barge Guide.

- Isolating valves are to be provided and located to permit optimum utilisation in the event of physical damage to any part of the main.
- No connections other than those necessary for firefighting purposes are allowed.
- Precautions against freezing must be taken as per 3/5.3.1 of the Accommodation Barge Guide.
- Materials readily rendered ineffective by heat are not to be used for fire mains unless adequately protected. The pipes and hydrants are to be in locations which are easily accessible and fire hoses may be easily coupled to them.
- A cock valve should be fitted to serve each fire hose so that any fire hose may be removed while the fire pumps are at work.
- The location of pumps, sea suction, and power sources for fire pumps are to be such that a fire or other casualty in one space will not render both main pumps inoperable.

Hydrant locations and arrangements

The number and position of hydrants should be such that at least two jets of water, not emanating from the same hydrant (one of which should be from a single length of fire hose), may reach any part of the unit normally accessible to those onboard. A hose should be provided for every hydrant. Fire hoses are to be of material approved and be sufficient in length so the jet of water can reach any space for which it is intended to be used. The hose length must not exceed 23 m (75 ft). Dual-purpose nozzles and necessary couplings are to be provided as per 3/5.3.1 of the Accommodation Barge Guide. Any necessary fittings and tools must be kept ready for use in conspicuous positions near the hydrants or connections. At interior locations, fire hoses are to be connected to the hydrants at all times.

Nozzles

Standard nozzle sizes are to be 12 mm (0.5 in), 16 mm (0.625 in), and 19 mm (0.75 in), or as near to as possible. For accommodation and service spaces, a nozzle size greater than 12 mm (0.5 in) is not required. For machinery spaces and exterior locations, the nozzle size should be such as to obtain the maximum discharge possible from two jets at the referenced pressures in 3/5.3.1 of the Accommodation Barge Guide from the smallest pump. However, a nozzle size greater than 19 mm (0.75 in) is not required.

International shore connection

The barge should be provided with at least one international shore connection complying with SOLAS regulation II-2/10.2.1.7. Facilities should be available enabling such a connection to be used on any side of the barge.

OFFSHORE SUPPORT VESSELS

Revised notation

The Class rules for offshore support vessels are essentially derived from the same or similar rules for building and classing steel vessels.

Emergency fire pump suction

The text 'In selecting the emergency fire pump, the minimum available net positive suction head should provide a safety margin of at least 1 m (3.3 ft) or 30% of required net positive suction head of the pump, whichever is less' in Chapter 3 should be ignored as offshore support vessels are not required to provide a safety margin for the minimum number of emergency fire pumps.

In this chapter, we have discussed the main Class rules as they relate to the classification of shipboard fire main systems. Although the guidance in this chapter is sufficient to provide an overview of universal Class requirements, reference should always be made to the specific Class rules pertaining to the vessel in question.

Chapter 5

Fixed water extinguishing systems

Water is an ideal extinguishing medium for many shipboard applications. It is readily available, has excellent heat-absorbing capabilities, and can be used on a variety of fires. There are several mechanisms involved in the extinguishing effect of a fire with water. In this chapter, we will discuss those mechanisms, the advantages of using water as an extinguishing media, and the limitations of water as an extinguishing media. First, there is the cooling of the flame temperature when water passes through the combustion zone and absorbs heat through evaporation. Cooling of the flame temperature results in a reduction in the amount of radiant heat released by the fire, and therefore, a reduction in the amount of heat radiated back to the fuel surface. Second, there is the cooling effect of the fuel surface by the direct impingement of water droplets on the surface. With a reduction of the radiant heat received at the fuel surface and the additional cooling of the fuel surface by direct contact with the water droplets, there is a reduction in the number of combustible gases released. With sufficient cooling of the flame temperature and/or the fuel, the rate of pyrolysis or vapourisation of combustible vapours will be reduced to a point where combustion will no longer be self-supporting. Water has the important additional effect that when it evaporates, it turns into steam. The steam, which is in the immediate vicinity of the chemical reaction, displaces the air that supplies oxygen for the combustion process and results in a smothering of the fire.

Fixed water extinguishing systems are normally considered to include water spray, water sprinkler, and water mist systems. These systems utilise fixed piping systems with distributed arrays of nozzles located in the overhead, which are supplied from dedicated pump(s). However, the particular fire hazards and safety concerns vary depending on the particular type of space being protected. For example, in a machinery space, one would anticipate Class 'B' combustibles to be involved, while in an accommodation space, one would anticipate the involvement of Class 'A' combustibles. Even the degree of anticipated supervision has a role. There are many locations in the accommodation spaces and service spaces that are not continuously supervised (cabins, storage cupboards, etc.) and a small initial fire could easily go

DOI: 10.1201/9781003385523-7

unnoticed by shipboard personnel. There are also certain differences in the extinguishing mechanisms at work for a water mist system as compared to those involved in a water spray or water sprinkler system. Accordingly, the system designs, as well as the requirements, vary depending upon the space to be protected and the type of system to be installed. The following provides a brief discussion regarding the individual types of systems.

GENERAL PRINCIPLES OF FIXED WATER EXTINGUISHING SYSTEMS

Water spray systems

Water spray systems are manually operated, open 'deluge'-type systems, and are typically used to protect open RORO spaces, 'Special Category' spaces, and cargo pump rooms. Water spray systems are also permitted for use in protecting machinery spaces, but such arrangements are seldom proposed due to the hazards involved and the availability of suitable alternatives. Like other types of fixed water extinguishing systems, the water spray system consists of pump(s), a fixed piping system, and a distributed array of nozzles. However, the one distinctive characteristic of a water spray system is that it utilises 'open'-type nozzles. When this type of system is activated, water will be discharged simultaneously through all of the branch nozzles, which can result in very high water demand rates. Also, these systems are normally required to be manually actuated, and therefore personnel must initiate the discharge of the system.

Water sprinkler systems

Water sprinkler systems also utilise a distribution system consisting of fixed supply piping and overhead nozzles. However, unlike water spray systems, water sprinkler systems are designed for automatic activation. Since these systems are automatically actuated, the distribution system must be always pressurised. To accommodate the automatic release, the sprinkler nozzles used are of the 'closed' type and fitted with individual heat-sensitive links or bulbs that allow the nozzle to open when the temperature of the air in the vicinity of a particular nozzle exceeds a certain 'activation' temperature. Since each individual nozzle has its own activation mechanism, only those nozzles in the immediate vicinity of the fire will see temperatures sufficient for activation and will open. Normally, the nozzles located directly above or next to the fire source, as well as the nozzles around the outside perimeter of the fire base, will be opened. Those nozzles located directly over or near the fire source serve to control the fire by wetting the flames and fuel source, while the nozzles around the perimeter of the fire serve to pre-cool any surrounding combustible materials. These systems are normally used to protect accommodation spaces and service spaces.

Water mist systems

Water mist technology has been around for some time. However, only in recent years have appropriate design and testing criteria been developed by the IMO for water mist systems in marine applications, permitting their consideration as acceptable alternatives to the required water spray and sprinkler systems. Water mist systems have pumping and distribution piping system arrangements like those found with water spray and water sprinkler systems. However, the critical difference between this type of system and either a water sprinkler or water spray system is in the size of the water droplets discharged from the nozzles. The water droplets associated with a water mist system are typically much finer than those discharged from either a water spray or water sprinkler nozzle. For some water mist systems generating the smaller droplet sizes, the extinguishing medium discharged is like a thick fog. Several different techniques are used to generate the smaller water mist droplets, including impingement, pressure jet, and twin fluid methods. The impingement method typically involves directing the water onto a spiral or spinning device, which breaks up the stream into small, well-distributed droplets. The pressure jet method involves the acceleration and dispersion of small water droplets through swirling chambers or high-pressure discharges through very small orifices. The twin fluid method atomises the water by injecting the water jet and an atomising gas together in such a way as to shear the water droplets into smaller particles, creating a mist. Depending upon the manufacturer's design, water mist system pressures can range from very low pressure, 3–5 bar (43–72 psi) to quite high pressures, 100–150 bar (1,450–2,175 psi).

The actual size of the water mist droplets created will vary depending upon the technique used and the degree of nozzle refinement developed by the individual manufacturer. It is important to note that the actual size of the water droplet will impact the degree to which the various extinguishing mechanisms discussed in the first paragraph of this section will be actively involved. To explore this issue further, one must note that water spray and sprinkler systems discharge relatively 'large' water droplets. Due to the weight and/or inertia of these 'large' droplets, they 'fall' into the flames. As these 'large' droplets pass through the flame, a certain amount of the droplet will evaporate. The evaporation absorbs heat from the flame, which results in a cooling of the flame temperature. A certain amount of steam is also generated. However, due to the 'large' size, often a significant portion of the droplet will pass through the flame intact and then contact the fuel surface. The cooling of the fuel surface by the water droplet plays a major role in the control or extinguishing effect of the fire by a water spray or water sprinkler system.

In contrast to water spray or water sprinkler droplets, water mist droplets are typically much smaller. The reduced size of the droplet can significantly impact the extinguishing mechanisms at work. First, the smaller water mist droplets have minimal weight, and their flow paths are more strongly affected by the fire plume convection currents. As a result, the water mist droplets do not typically pass through the flame due to their own weight and momentum, and in fact, may have difficulty reaching the flame due to the plume convection currents. Typically, the water mist droplets are carried into the fire with the air currents feeding the fire. These air currents normally enter the flame at the lower levels of the plume and the water mist droplets are pulled directly into the combustion zone with the air. This process provides a good distribution of the mist droplets within the combustion zone of the fire so long as the droplets are well entrained in the air currents feeding the fire. The second impact on the extinguishing mechanisms is that since the water droplets are much smaller, a much greater surface area per volume of water is presented to the fire. This facilitates greater heat absorption and a higher evaporation rate for a particular volume of water. That is, for the volume of water actually entering in the flame region, the smaller water mist droplets will typically absorb more heat than the typical water spray or water sprinkler droplets due to the increased amount of exposed surface area. This results in a greater degree of cooling for the flame per unit volume of water, and therefore, a greater reduction in the amount of heat radiated back to fuel surface. This increase in heat absorption also increases the amount of evaporation and produces greater steam generation, and as a result, the smothering action of the fire by steam is increased. There are, however, certain negative impacts associated with the reduced size of the water mist droplets. For spray and water sprinkler systems, the cooling of the fuel surface, especially in the immediate vicinity of the base of the fire, can play a very important role in the control and extinguishing effect of a fire. However, since the water mist droplets are not as large as the typical water spray or water sprinkler droplets, fewer of the water mist droplets are able to pass through the flame and impact the fuel surface within the base of the fire while still intact. Also, the lack of weight or significant inertia of the water mist droplet makes it more difficult to ensure that water droplets will follow a predictable path. Therefore, more of a 'total flooding' effect is needed for spaces being protected with water mist applications utilising smaller size droplets. Water mist systems have been proposed for use in machinery spaces, as well as in accommodation and service spaces. To address such proposals, IMO developed fire test criteria for establishing the equivalence of a particular water mist manufacturer's system to the water spray addressed in the FSS Code, Chapter 7, and the water sprinkler systems addressed in the FSS Code, Chapter 8. IMO MSC/Circ.1165, as amended by IMO MSC.1/Circs.1237 and 1269 and associated with IMO MSC.1/Circs.1385 and 1386, provides the testing criteria applicable to establish equivalence for use in machinery spaces and cargo pump rooms, while IMO Resolution A.800(19), as amended by MSC.265(84) and MSC.284(86), provides the testing criteria applicable to establish equivalence to a water mist system for use in accommodations and service spaces.

CLASS REQUIREMENTS FOR FIXED WATER SPRAY, WATER SPRINKLER, AND WATER MIST SYSTEMS

The following information is provided as general guidance on the Class requirements for water spray, water sprinkler, and water mist systems. However, reference should always be made to the Class rules applicable to the specific vessel concerned for the complete set of requirements.

General system component requirements

Pumps

Pumps associated with water sprinkler, water spray, or water mist systems are considered to be the same as fire pumps and are required to be Class certified.

Piping components and materials

System piping components are to comply with the general requirements and limitations set by Class. In addition, depending upon the pressure 'Class' of the system, the components may also be required to be certified in accordance with additional Class requirements. A submitted drawing should include a complete Bill of Materials that provides the material specifications, standards of construction, pressure ratings, and types of all valves and fittings, as well as the material specification, sizes, and wall thickness of the piping. The drawing should also identify the maximum system pressure and provide adequate information to identify the layout and arrangement of the system(s). In addition to common requirements and limitations set by Class, the materials used in the system are not to be rendered ineffective by heat. To be considered not 'readily rendered ineffective by heat', a component should be certified as having passed an acceptable recognised fire test, or the material should have a melting temperature higher than the test temperature specified in an acceptable fire test. Resiliently seated valves may also be considered for use in a water spray, water sprinkler, or water mist system, provided that the proposed valves can pass an appropriate fire test acceptable to Class and can be effectively closed with the resilient seat damaged or destroyed such that leakage through the closed valve is insignificant. However, rubber (cloth inserted) gaskets are not only frequently proposed but also not permitted, as their destruction under the fire could lead to the loss of system pressure and integrity. In addition, particular attention should be exercised regarding the arrangements of sea suctions, pumps' suction, and discharge outlets for compliance with Class rules and the arrangements required at the firefighting station for fixed fire extinguishing systems in the propulsion machinery space. For regulatory compliance, criteria stipulated by the Flag administration must be satisfied and may differ from the criteria stated by Class.

Nozzles

The nozzles are a very critical component of any fixed pressure water spray, water sprinkler, or water mist system, and their proper operation is crucial to the system's ability to control and/or extinguish a fire. Accordingly, the performance of nozzles should be verified through testing by an appropriate laboratory, agency, or other recognised entity (i.e. Flag State, Administration, and Class) to a suitable standard or procedure, which is specifically designed to establish and verify the ability of the nozzle design to consistently provide the discharge spray density, droplet size, coverage pattern, etc., as advertised by the manufacturer. In addition to the above, Class may also require all fixed fire extinguishing system components, which would include spray, sprinkler, or mist system nozzles, to be covered under the relevant Class Certification Programme. Accordingly, such nozzles must also meet the requirements specified in the table for either an 'Individual Unit Certification' or the Class Type Approval Programme, as appropriate.

Fixed water spray systems in machinery spaces

The Class rules recognise pressurised water spray systems as acceptable fixed fire extinguishing arrangements for use in machinery spaces. The requirements for fixed water spray systems in machinery spaces can be found in the FSS Code in Chapter 7/2.1 and are discussed below.

Nozzles

FSS Code, Chapter 7/2.1.1.1, requires any required fixed pressure water-spraying fire extinguishing system in machinery spaces to be fitted with spraying nozzles of an approved type. As discussed earlier, the approval of the nozzle by an appropriate laboratory or agency is very important in documenting the nozzle's performance and suitability for the intended service. Documentation verifying such approvals should be available.

Coverage rate

FSS Code, Chapter 7/2.1.1.2, requires the number and arrangement of the water spray system nozzles to be sufficient to provide an effective average distribution of water of at least 5 l/min/m^2 (0.12 gpm/ft^2) over the protected area. FSS Code, Chapter 7/2.1.2.1, further requires nozzles to be fitted above bilges, tank tops, and other areas over which oil fuel is liable to spread, as well as other specific fire hazards in machinery spaces. In reviewing the arrangement of the nozzles for a particular application, the spray density, recommended spacing, and coverage pattern identified by the manufacturer for a particular nozzle should be compared to the proposed layout and spacing. The review must verify that the design will provide the required spray coverage rate to all portions of the protected space, as well as verify nozzles are fitted above bilges, tank tops, and other areas over which oil fuel is liable to spread, as well as other specific fire hazards in machinery spaces. It should be noted that FSS Code, Chapter 7, also makes mention of the possible need to require increased application rates, where considered necessary by the Flag administration.

Related IMO MSC.Circ.1120 Interpretation.

SOLAS II-2/10.2 (FSS Code, Chapter 7) Areas for increased application rates.

An indication of areas for which increased application rates may be required is given below:

In reviewing the nozzle layout and spacing, it is a very important to recognise that the spray density and distribution pattern for a particular nozzle are functions of the nozzle height, as well as the water pressure at the nozzle inlet. While the height of the nozzles should be identified in the drawing, pressure drop calculations are normally necessary to establish pressure at the most hydraulically remote nozzle section. While it may be possible to calculate the system pressure drops for a simple 'tree' water spray system, more elaborate tree systems, as well as 'grid' or 'looped' distribution piping systems, are

Table 5.1 Coverage and Application Rates for Water Spray Systems

Protected area	Application rate
Boiler fronts or roof, firing areas, oil fuel units, centrifugal separators (not oily water separators), oil purifiers, and clarifiers	20 l/m
Hot oil fuel pipes near exhausts or similar heated surfaces on main or auxiliary diesel engines	10 l/m

normally used for larger water spray systems and pressure drop calculations for these systems become very complicated. Accordingly, the submission of pressure drop calculations from the designer is often necessary for all but the simplest of designs. Once received, the pressure drop calculations should be reviewed for the analytical approach used, the initial criteria and values (e.g. friction factors, flow rates, and pipe sizes), and overall completeness of the analysis and the results. With more complex systems, most analyses will be computer-based. Without the actual programme, it is often very difficult to completely verify all portions of such calculations. Therefore, additional information to substantiate the validity of the computer programme is also frequently necessary.

Division into sections

Water spray systems in machinery spaces can be quite large. Accordingly, FSS Code Chapter 7/2.1.2.2, allows the system to be divided into sections. The distribution valves controlling the various sections are required to be operated from easily accessible positions outside the spaces to be protected and in a location which is not readily cut off by a fire in the protected space.

Supply and control arrangements

Quick and ready operation is vital for any firefighting system, and especially one which is intended to protect a space containing the volume of combustibles normally found in a machinery space. Accordingly, FSS Code, Chapter 7/2.1.3, requires the water spray system to be kept charged at the necessary pressure, and the pump supplying the water for the system should be put automatically into action by a pressure drop in the system. There are several points to consider with this requirement. First, the system will have open-type nozzles, and therefore, the pressurisation of the system will only be up to the zone valves. Second, the type of system selected to maintain the continuous pressure in the water spray main is important. The continuous pressurisation of the system is normally provided either through an accumulator system or by a pump. However, if a pump is used, it is important to verify that its design will be appropriate for this service. Main system pumps are usually designed for large capacities and are not suitable for this pressure maintenance duty, since continuous operation at shut-off head will damage the pump. However, smaller capacity 'jockey' pumps, which cycle on and off to maintain the pressure in the water spray main, can be used for this type of service. The third item to note is that the valve arrangement, on both the supply and discharge sides of the main pump, must be arranged so that the appropriate valves will be open when the water spray pump is automatically brought into action so that the system will operate properly.

System capacity requirements

FSS Code, Chapter 7/2.1.1.4, requires that the water spray pump be capable of simultaneously supplying at the necessary pressure all sections of the system in any one compartment to be protected. Since the nozzles are of the open type, all nozzles of a particular branch or zone will be brought into action when that branch or zone valve is opened. When considering all branches or zones within a particular machinery space, the result can be a substantial demand on the pumping and distribution system capacity. System flow and pressure drop calculations, as discussed earlier, are necessary to verify that the pump and distribution system comply with this requirement.

Arrangements

FSS Code, Chapter 7/2.1.2.3, requires that the pump and its controls are installed outside the space or spaces to be protected and that the arrangements are such that a fire in the space or spaces protected by the water spray system will not render the system inoperative. Accordingly, the arrangements for the pump location, the power supply to the pump and controls, access and ventilation arrangements, etc., for the water spray system, must be examined to confirm compliance.

Pump prime movers

FSS Code, Chapter 7/2.1.1.5, addresses requirements for the prime mover of the system pump and indicates that the pump may be driven by independent internal combustion machinery. However, if the prime mover for the pump is dependent upon power being supplied from the emergency generator, the generator is required to be arranged to start automatically in the case of main power failure so that the power for the pump is immediately available. Further to the requirements outlined in FSS Code, Chapter 7/2.1.2.3, the requirements in FSS Code, Chapter 7/2.1.1.5, specify that if the pump is driven by independent internal combustion machinery, then the arrangements are to be such that a fire in the protected space will not affect the air supply to the machinery.

Protection from clogging

FSS Code, Chapter 7/2.1.1.3, requires precautions to be taken to prevent the nozzles from becoming clogged by impurities in the water or corrosion of piping, nozzles, valves, and pump. This is normally accomplished using galvanised piping, in-line strainers, and flushing and/or drainage arrangements.

Fixed water sprinkler systems in accommodation spaces

For all vessels utilising Method 'IIC' construction, Class rules require an automatic sprinkler system to be fitted in the accommodation spaces, galleys, and other service areas. In addition, there is a requirement that all accommodations and service spaces on passenger vessels should be fitted with an automatic sprinkler system unless such spaces are fitted with an approved fixed fire detection and fire alarm system. The system requirements for the automatic sprinkler systems installed in the above spaces can be found in the appropriate Class rules and are in line with the requirements found in FSS Code, Chapter 8. These system requirements are discussed below.

General

The Class rules indicate that any required automatic sprinkler system should be capable of immediate operation at all times and that no action by the crew should be necessary to set it in operation. Moreover, it is typically stipulated in the Class rules that the system should be of the wet pipe type (i.e. permanently filled with water under pressure from the freshwater tank), should be kept charged at the necessary pressure, and should have provision for a continuous supply of water. The Class rules do permit small, exposed sections of the system to be of the dry pipe type (i.e. not permanently filled with water) where this is a necessary precaution. Saunas should be fitted with a dry pipe sprinkler system. The operating temperature may be up to 140 °C (284 °F). There is a general requirement that any parts of the system, which may be subjected to freezing temperatures in service, are to be suitably protected against freezing.

Visual and audible alarms

Each section of sprinklers should include the means for giving a visual and audible alarm signal automatically at one or more indicating units whenever any sprinkler comes into operation. The alarm systems are also capable of indicating any fault occurring in the system. Such units are to indicate in which section served by the system a fire has occurred and are to be centralised on the navigation bridge or in the continuously manned central control station. In addition, visible and audible alarms from the unit are to be in a location other than in the aforementioned spaces so that the indication of fire is immediately received by the crew.

Zone arrangements

Sprinklers are required to be grouped into separate sections, each of which should contain no more than 200 sprinklers. In this way, damage to piping within a particular zone or maloperation of a particular zone valve will not disable the entire sprinkler system. It also requires that in passenger vessels, any section of sprinklers is not to serve more than two decks and is not to be situated in more than one main vertical zone. However, Class may permit such a section of sprinklers to serve more than two decks or be situated in more than one main vertical zone if it is satisfied that the protection of the vessel against fire will not thereby be reduced.

Section isolation valves

The Class rules provide requirements for section isolating valves and require each section of sprinklers to be capable of being isolated by one stop valve only. The stop valve in each section should be readily accessible in a location outside of the associated section or in cabinets within stairway enclosures, and its location should be clearly and permanently indicated. Means are to be provided to prevent the operation of the stop valves by any unauthorised person.

Pressure gauges

Pressure gauges, capable of indicating the pressure in the system, must be provided at each section stop valve and at a central station. This allows the crew to verify that the system is pressurised and ready for operation.

Zone arrangement plan

To assist the crew in identifying the location of a fire quickly, the Class rules require a list or plan to be displayed at each indicating unit showing the spaces covered and the location of the zone in respect of each section. Suitable instructions for testing and maintenance are also required to be available.

Nozzles

In addition to the general requirements for nozzles, most Class rules specifically indicate that the sprinkler nozzles are to be resistant to corrosion by the marine atmosphere. Approval by an appropriate laboratory or agency for use in marine applications would be sufficient to verify compliance with this requirement. Within accommodation and service spaces, the sprinklers must come into operation within the temperature range from 68 °C (154 °F) to 79 °C (174 °F). The one stated exception to this requirement is in locations such as drying rooms, where high ambient temperatures might be expected. For these applications, the operating temperature may be increased by not more than 30 °C (86 °F) above the maximum deck head temperature. The design, selection, and proper operation of the sprinkler nozzles are critical to the effectiveness with which an automatic sprinkler system can control or extinguish a fire. The sprinkler nozzles must be located and installed in accordance with the manufacturer's recommendations. Certain nozzles are designed to be located near a side wall, while others are designed for a full pattern. In addition, there are upright, as well as pendent, sprinkler nozzles. As discussed in Chapter 5, the coverage pattern is influenced by the actual nozzle height and the pressure at the nozzle. Accordingly, such details and information must be considered when reviewing the placement of the nozzles.

Coverage rate

Sprinklers must be placed in an overhead position and spaced in a suitable pattern to maintain an average application rate of not less than 5 l/min/m² (0.12 gpm/ft²) over the nominal area (i.e. gross, horizontal projection of the area to be covered) covered by the sprinklers.

Related IMO MSC Circ.1120 Interpretation.

SOLAS II-2/12.3 (FSS Code, Chapter 8/2.5.2.3) 'Definition of nominal area'.

Nominal area is defined as being the gross, horizontal projection of the area to be covered.

Automatic operation

The system must be capable of automatic operation in the event of a fire. So that an adequate quantity of water at the necessary pressure is available at all times, the Class rules require the system to be fitted with a freshwater pressure tank. The tank should contain a standing charge of fresh water equivalent to the amount of water which would be discharged in one minute by the pump, and the arrangements are to provide for maintaining an air pressure in the tank such that when the standing charge of fresh water in the tank has been used, the pressure will be not less than the working pressure of the sprinkler, plus the pressure exerted by a head of water measured from the bottom of the tank to the highest sprinkler in the system. To accommodate the volume of air necessary to maintain the pressure throughout the one-minute discharge, the tank is required to have an actual volume equal to at least twice that of the charge of water specified earlier. In addition, suitable means of replenishing the air under pressure and of replenishing the freshwater charge in the tank are to be provided. Also, a glass gauge is required to be provided to indicate the correct level of the water in the tank.

Accumulator

The accumulator should be considered as a pressure vessel, and it should comply with the applicable requirements stipulated by Class. These requirements will vary depending upon the maximum pressure involved. During the review of the system, the applicable requirements set by Class should be identified and the submitter advised accordingly unless Class certification is already indicated on one of the drawing notes. To prevent corrosion and the possibility of excessive sedimentation, Class may require that a means be provided to prevent the passage of seawater into the tank.

Sprinkler pump

Automatic operation
An independent power pump must be provided solely for the purpose of continuing the automatic discharge of water from the sprinklers. The pump should be brought into action automatically by the pressure drop in the system before the standing freshwater charge in the pressure tank is completely exhausted. The system's control arrangement should be reviewed to verify compliance with this requirement.

Pump performance
The pump and the piping system are to be capable of maintaining the necessary pressure at the level of the highest sprinkler to provide a continuous output of water sufficient for the simultaneous coverage of a minimum area of 280 m² (3,050 ft²) at the application rate specified in the appropriate Class rules. The ability of the pump and the piping system to support this flow rate at the most remote section must be verified through a review of the pump curves and appropriate pressure drop calculations, as discussed in Chapter 5.

Pump testing arrangements
A test valve with a short open-ended discharge pipe should be fitted on the delivery side of the pump. The effective area through the valve and pipe is required to be adequate to permit the release of the required pump output while maintaining the pressure in the system.

Sea inlet
The sea inlet to the pump should be in the space containing the pump wherever possible. In addition, the arrangement should be such that when the vessel is afloat, it will not be necessary to shut off the supply of sea water to the pump for any purpose other than the inspection or repair of the pump.

Location
To protect the integrity of the sprinkler system, the Class rules require that the sprinkler pump and tank be located in a position reasonably remote from any machinery space of Category 'A', since Category 'A' machinery

spaces are high fire risk areas. It further requires that the sprinkler pump and tank are not to be in any space that is required to be protected by the sprinkler system.

Source of power

In accordance Class rules, there should be no less than two sources of power supply for the seawater sprinkler pump and automatic alarm and detection system. If the pump is electrically driven, it is required to be connected to the main source of electrical power, which should be capable of being supplied by at least two generators. Furthermore, for passenger vessels, an electric motor powering the sprinkler pump should also be supplied with a feeder from the emergency switchboard. The feeders are to be arranged so as to avoid galleys, machinery spaces, and other enclosed spaces of high fire risk, except insofar as it is necessary to reach the appropriate switchboards. One of the sources of power supply for the alarm and detection system should be an emergency source. Where one of the sources of power for the pump is an internal combustion engine, in addition to complying with the provisions set by Class, the source should be so situated that a fire in any protected space will not affect the air supply to the machinery.

Back-up supply of water

To provide a backup source of water, the Class rules require that the sprinkler system must have a connection from the vessel's fire main by way of a lockable screw-down non-return valve which will prevent a backflow from the sprinkler system to the fire main. While the interconnection for a backup source of water is required, this does not necessarily require the fire main pumps to be sized to supply the sprinkler system or the fire main and sprinkler system simultaneously.

System testing arrangements

Testing of the system is very important in confirming that the sprinkler system will operate properly when needed. Accordingly, the Class rules require a test valve to be provided for testing the automatic alarm for each section of sprinklers by a discharge of water equivalent to the operation of one sprinkler. The test valve is required to be situated near the stop valve for that section. In addition, a means for testing that the pump will operate automatically upon reduction of pressure in the system should be provided. Switches are also required to be provided at one of the indicating positions referred to by Class which will enable the alarm and the indicators for each section of sprinklers to be tested.

Spares

As previously discussed, the sprinkler nozzles are of the 'closed' type, and each individual nozzle is fitted with its own heat-sensitive link or bulb, which allows the nozzle to open when exposed to excessive heat. Accordingly, if one sprinkler head activation link or bulb breaks or otherwise allows the sprinkler head to open inadvertently, the entire zone must be shut down until the sprinkler head is replaced. Thus, it is very important to carry an adequate number of spare sprinkler heads onboard so that a defective head can be replaced and the particular sprinkler zone re-pressurised as soon as possible. It is also important to recognise that different locations are fitted with different types of heads or activation temperatures. The Class rules will typically stipulate the requirement that spare sprinkler heads are to be provided for each section of sprinklers. Spare sprinkler heads should include all types and ratings installed in the vessel, and should be provided as follows:

- < 300 sprinkler heads   6 spare sprinkler heads
- ~ 300 to 1,000 sprinkler heads   12 spare sprinkler heads
- > 1,000 sprinkler heads   24 spare sprinkler heads

The number of spare sprinkler heads of any type need not exceed the number of heads installed of that type.

Fixed water spray systems in RORO spaces

The protection of RORO spaces presents a unique challenge. There are frequently low flashpoint fuels (i.e. petrol) in vehicle fuel tanks, large quantities of Category 'A'-type combustibles, and the possibility of passengers in the area on ferry-type passenger vessels. However, to better understand the challenges, the differing requirements of where fixed water spray systems come into use, it is necessary to first understand several important definitions. These are provided as follows:

- *RORO cargo spaces.* This refers to any space(s) not normally subdivided in any way and which extend to either a substantial length or the entire length of the vessel, and in which goods [packaged or in bulk, in or on road cars, vehicles (including road tankers), trailer containers, pallets, demountable tanks or in or on similar stowage units or other receptacles] can be loaded and unloaded normally in a horizontal direction.
- *Open RORO cargo spaces.* RORO cargo spaces refers to those which are open at both ends or open at one end and provided with adequate natural ventilation effective over the entire length through permanent

openings in the side plating or deck head to the satisfaction of the administration.

- *Closed RORO cargo spaces.* This applies to RORO cargo spaces which are neither open RORO cargo spaces nor weather decks.
- *Vehicle passenger ferry.* A vessel designed and fitted for the carriage of vehicles and more than 12 passengers.
- *Special category spaces.* Special category spaces are those enclosed spaces above or below the bulkhead deck intended for the carriage of motor vehicles with fuel in their tanks for their own propulsion, into and from which such vehicles can be driven and to which passengers have access.

Notwithstanding these definitions, two further distinctions should be made in reference to RORO spaces. These are RORO spaces on cargo vessels and passenger vessels.

RORO spaces on cargo vessels

The fire extinguishing arrangements for a typical RORO space on a cargo (i.e. non-passenger) vessel are addressed in the relevant Class rules for that vessel. Typically, the Class rules require RORO spaces which are capable of being sealed (i.e. closed RORO space) to be fitted with either a fixed gas fire extinguishing system or a fixed pressure water spray system. However, for RORO spaces on such vessels that are not capable of being sealed (i.e. open RORO spaces), a fixed gas system would be deemed totally ineffective as a concentration of gas cannot be maintained, and therefore, a fixed pressure water spray system is required. For the requirements of the fixed pressure water spray system, refer to SOLAS regulation II-2/20.6.1.2, FSS Code, IMO MSC.1/Circ.1272 *Guidelines for the approval of fixed water-based firefighting systems for RORO spaces* and *special category spaces equivalent to that referred to in Resolution A.123(V)* and IMO MSC.1/Circ.1430 *Revised guidelines for the design and approval of fixed water-based firefighting systems for RORO spaces and special category spaces.*

RORO spaces on passenger vessels

As noted earlier, ferries present the additional concern of possibly having passengers in the RORO space when activation of the fire extinguishing system is needed. Therefore, SOLAS and Class rules differentiate between those RORO spaces where passengers are permitted access and those in which passengers are not. As defined earlier, those RORO spaces to which passengers may have access are called 'Special Category Spaces', and Class rules require that all such spaces be fitted with a pressure water spray system complying with SOLAS regulation II-2/20.6.1.2 and the FSS Code, for which the use of any other fixed fire extinguishing system instead may be permitted in accordance with SOLAS regulation II-2/20.6.1.3 referring to IMO MSC.1/Circ.1272 and IMO MSC.1/Circ.1430. For those RORO spaces on passenger vessels which are not accessible by passengers, SOLAS regulation II-2/20.6.1.1 permits the use of fixed fire extinguishing systems similar to those permitted in cargo vessel RORO spaces (for instance, fixed gas or fixed water spray systems). As indicated earlier, fixed pressure water spray systems installed in RORO spaces must comply with the FSS Code; the detailed requirements of which have been discussed in Chapter 5. Another important issue regarding water spray systems is the recognition that these types of systems can discharge large quantities of water into protected spaces in a relatively short period of time. This additional weight can have seriously deleterious effect on the vessel, especially if large quantities of water are introduced at higher levels within the vessel's structure. The effect of the weight of this water should raise the centre of gravity of the vessel with the resulting free surface effect detrimentally affecting the stability of the vessel, especially in large open areas, such as RORO decks, where the potential for large free surface effect exists. Accordingly, SOLAS regulation II-2/20.6.1.4 requires that drainage and pumping arrangements are to be such as to prevent the build-up of free surfaces. If this is not possible, the adverse effect upon the stability of the added weight and free surface of water must be accounted for in the Class approval of the stability information. This is an issue that must be addressed early on so that potential problems with drainage or stability are avoided.

Fixed water mist systems in machinery spaces and cargo pump rooms

With the prohibition of Halon-based extinguishing systems and the hazards associated with CO_2 systems, as well as water spray systems, in machinery spaces and cargo pump rooms, the IMO has recognised the need for the consideration of alternative fire extinguishing systems. For this reason, the IMO developed MSC/Circ.1165, as amended by IMO MSC.1/Circs.1237 and 1269 and associated with IMO MSC.1/Circs.1385 and 1386, which provide guidelines for the approval of alternative water-based fire extinguishing systems as equivalent to the water spray systems referred to in FSS Code, Chapter 7. It is through compliance with the design and testing requirements specified in IMO MSC/Circ.1165, as amended by IMO MSC.1/Circs.1237 and 1269 and associated with IMO MSC.1/Circs.1385 and 1386, that manufacturers of water mist systems can establish that their system provides an equivalent level of safety, and therefore, an acceptable alternative to the water spray system discussed in FSS Code, Chapter 7.

Paragraph one of the annex to MSC/Circ.1165, as amended by IMO MSC.1/
Circs.1237 and 1269, provides a general explanation of the purpose for
the Circular. Paragraphs two through nine of the annex to MSC/Circ.1165,
as amended by IMO MSC.1/Circs.1237 and 1269, provide a number of
definitions associated with water-based fire extinguishing system, and paragraphs ten through 27 of the annex to MSC/Circ.1165, as amended by IMO
MSC.1/Circs.1237 and 1269, identify the general requirements for water
mist systems. Appendix 'A' to MSC/Circ.1165, as amended by IMO MSC.1/
Circs.1237 and 1269, identifies the design and testing requirements applicable to water mist nozzles and Appendix 'B' to MSC/Circ.1165, as amended
by IMO MSC.1/Circs.1237 and 1269, identifies the fire tests which must be
successfully completed for a manufacturer to qualify their system. A copy
of MSC/Circ.1165, as amended by IMO MSC.1/Circs.1237 and 1269 and
associated with IMO MSC.1/Circs.1385 and 1386, should be available for
any review involving a fixed water mist system to be installed in machinery
space and cargo pump room.

The following provides a discussion regarding the requirements applicable
to water mist systems installed in machinery spaces and cargo pump rooms
based on the requirements of IMO MSC/Circ.1165, as amended by IMO
MSC.1/Circs.1237 and 1269 and associated with IMO MSC.1/Circs.1385
and 1386.

(1)   General
    Water mist systems are specialised systems, and the characteristics
    of the systems (e.g., the required minimum pressure, droplet size and
    velocity, and minimum required discharge flow rate) will most likely
    vary substantially from manufacturer to manufacturer. As outlined
    in the general discussion earlier, there are several approaches used
    to generate water mist. Some manufacturers use high pressure sys-
    tems, whilst others use low pressure; some use mechanical impinge-
    ment, while others use jet discharge or atomisation, and so forth.
    The design of the various systems' nozzles differs dramatically
    depending on the approach taken, and they are specially designed to
    maximise the manufacturer's particular method of mist generation.
    Accordingly, the nozzles are particularly critical to the proper opera-
    tion and effectiveness of a water mist system. As indicated earlier,
    fixed pressure water spray systems installed in RORO spaces must
    comply with the FSS Code, the detailed requirements of which have
    been discussed throughout this Chapter 5. Another important issue
    regarding water spray systems is the recognition that these types
    of systems can discharge large quantities of water into protected
    spaces in a relatively short period of time. This additional weight
    can seriously affect the stability of vessels, especially when large

quantities of water are introduced at higher levels within the vessel's structure. The weight of the water raises the centre of gravity of the vessel and the free surface effects can detrimentally affect the stability of the vessel, especially in large open areas, such as RORO decks, where the potential for large free surface effect exists. Accordingly, SOLAS regulation II-2/20.6.1.4 requires that drainage and pumping arrangements are to be such as to prevent the build-up of free surfaces. If this is not possible, the adverse effect upon the stability of the added weight and free surface of water must be accounted for in the Class approval of the stability information.

(2)   Water mist nozzle and system testing and approval
Recognising the critical role that the nozzles play, Appendix 'A' of MSC/ Circ.1165, as amended by IMO MSC.1/Circs.1237 and 1269, provides substantial design and testing requirements specifically for the water mist nozzles. The design criteria address numerous items regarding the nozzle, including body strength, threading, and flow constant value calculations, as well as specific tests, such as oven plunge, heat exposure, thermal shock, corrosion, impact, and clogging tests. In addition to the nozzle design and testing requirements specified in Appendix 'A', Appendix 'B' identifies a series of fire tests that are required to qualify the nozzle (in particular) and the system (in general) for suitability to extinguish a machinery space or cargo pump room fire. The fire test should be performed in a test apparatus consisting of:

- An engine mock-up of the size (width × length × height) of 1 m × 3 m × 3 m (3.2 ft × 9.8 ft × 9.8 ft) constructed of sheet steel with a nominal thickness of 5 mm (0.19 in). The mock-up is fitted with two steel tubes of 0.3 m (0.01 in) in diameter and 3 m (3.8 ft) in length that simulate exhaust manifolds and a grating. At the top of the mock-up, a 3 m2 tray is arranged (in accordance with Figure 1 of MSC/Circ.1165); and

- A floor plate system of the size (width × length × height) of 4 m × 6 m × 0.5 m (13.1 ft × 19.6 ft × 1.6 ft) surrounding the mock-up. Provision should be made for the placement of fuel trays, as described in Table 6.1 of MSC/Circ.1165 and located as described in Figure 6.1 of MSC/Circ.1165. The tests should be performed in a room having a specified area greater than 100 m2, a specified height of at least 5 m (16.4 ft), and ventilation through a door opening of 2 m × 2 m (6.5 ft × 6.5 ft) in size. The fire and engine mock-up should be according to Tables 1, 2, and 3 and Figure 6.2 of MSC/Circ.1165. The test hall should have an ambient temperature of between 10 °C and 30 °C (50 °F and 86 °F) at the start of each test.

As discussed previously, fixed fire extinguishing system components are required to be Class certified in accordance with the relevant Class rules. The review of the manufacturer's testing documentation for compliance is a very tedious and time-consuming undertaking. However, with the Class certification programme, there should only be the need to perform this review once for a particular manufacturer's system and nozzle. The results of the review and details pertaining to the system's design and installation criteria should be recorded on the Class certification report so that future reviewers will not have to repeat the task of reviewing the individual test results. The certification report should only be issued after verifying compliance with all design and testing requirements specified in Appendices 'A' and 'B' of IMO MSC/Circ.1165, as amended by IMO MSC.1/Circs.1237 and 1269. The report should clearly indicate complete details regarding the basis of the system's approval, including any details which could possibly affect the system's performance. As a minimum, these details should include:

- The system's design parameters (e.g., the size and class of machinery space for which it is approved, maximum enclosure height, models of nozzles, minimum nozzle pressures, nozzle flow rates particle size distribution, pumps and control valves used for the test, and maximum ventilation condition).
- Installation criteria used during the testing (e.g., nozzle spacing, minimum and maximum heights, and required standoff distances).
- Any other special issues or requirements which could potentially affect the satisfactory performance of future installations.

The recording of this information is critical since compliance with such details must be verified during the review of a specific water mist system installation. The certification report should also list the various system drawings, schematics, and other documentation used for the approval (complete with reference numbers, dates, etc.) so that it can be determined if future submissions are utilising the same system design. Further discussions regarding the detailed testing and approval procedures of the individual manufacturer's system and nozzles in Appendices 'A' and 'B' are beyond the scope of this discussion. However, it is important for the engineer conducting the individual system review for a particular installation to understand that there are substantial design and testing requirements applicable to the water mist system and nozzles. The manufacturer's system must pass a series of fire tests applicable to the specific size machinery space simply to qualify for consideration, as specified in paragraph

11 of MSC/Circ.1165, as amended by IMO MSC.1/Circs.1237 and 1269. Furthermore, the review of a particular system installation must verify compliance with all of the information/criteria included on the Class issued certification report.

(3) Release arrangements

In accordance with paragraph ten of MSC/Circ.1165, as amended by IMO MSC.1/Circs.1237 and 1269, the system must be capable of manual release. In addition, paragraph 21 of MSC/Circ.1165, as amended by IMO MSC.1/Circs.1237 and 1269, requires that the system's operational controls are to be available at easily accessible positions outside the spaces to be protected and are not to be liable to be cut off by a fire in the protected space(s). Verification of compliance with the above should either be verified during the plan review, otherwise a comment should be added requesting the Class Surveyor to verify the arrangements.

(4) Availability

The system should be available for immediate use and capable of continuously supplying water for at least 30 minutes to prevent re-ignition or fire spread within that same period of time. Systems that operate at a reduced discharge rate after the initial extinguishing period are to have a second full fire extinguishing capability available within a five-minute period of initial activation.

(5) Immediate and continuous supply

There are also several other points to consider with the requirements in paragraph 12 of MSC/Circ.1165, as amended by IMO MSC.1/Circs.1237 and 1269. First, for the water mist system to be 'available for immediate use and capable of continuously supplying water', the system must be kept charged at the necessary pressure and the pump supplying the water for the system must be arranged to come into action automatically by a drop of the pressure in the system. Second, the valving arrangement, on both the supply and discharge sides of the main pump, must be arranged such that the appropriate valves will be open when the sprinkler pump is automatically brought into action.

(6) Design and suitability of components

Paragraph 13 of MSC/Circ.1165, as amended by IMO MSC.1/Circs.1237 and 1269, requires that the system and its components are to be suitably designed to withstand ambient temperature changes, vibration, humidity, shock, impact, clogging, and corrosion normally encountered in machinery spaces or cargo pump rooms in vessels. Further, components within the protected spaces should be designed to withstand the elevated temperatures, which could occur during a fire. Compliance with these requirements should be verified during the review of the standards and approvals issued for the components during the review of the Bill of Materials. Paragraph 14 of MSC/Circ.1165, as amended by IMO MSC.1/Circs.1237 and 1269, requires that the system and its components 'be designed and installed in accordance with international standards acceptable to the Organisation and manufactured and tested to the satisfaction of the administration in accordance with the appropriate elements of Appendices A and B'. In this regard, a footnote is also provided which indicates the 'pending the development of international standards acceptable to the Organisation, national standards as prescribed by the Administration should be applied'. As one reviews the design and testing requirements identified in MSC/Circ.1165, as amended by IMO MSC.1/Circs.1237 and 1269 and associated with IMO MSC.1/Circs.1385 and 1386, it becomes obvious that the requirements are focused on the nozzle as well as some overall system requirements. However, there are many areas of a water mist system which are not adequately addressed in the Circular. To address such concerns, the authors of the Circular simply reference 'international standards acceptable to the Organisation'. However, at this time, no particular 'international standards acceptable to the Organisation' are in place, and therefore, the footnote allows consideration of appropriate national standards. While other appropriate national standards may exist, the one national standard which is frequently used is NFPA 750, *Standard on water mist fire protection systems*. Accordingly, based on the aforementioned reference, the engineer conducting the review should understand that in addition to compliance with the design and testing requirements specified in Appendices 'A' and 'B' of MSC/Circ.1165, as amended by IMO MSC.1/Circs.1237 and 1269, and associated with IMO MSC.1/Circs.1385 and 1386, the system, as well as its components, must also comply with the design and installation requirements specified in an appropriate national or international standard, such as NFPA 750. The review should verify compliance with all aspects accordingly.

(7)   Nozzle placement
In accordance with paragraph 15 of MSC/Circ.1165, as amended by IMO MSC.1/Circs.1237 and 1269, the nozzle location, type of nozzle, and nozzle characteristics are to be within the limits tested during the Appendix 'B' fire tests. This is part of the information which should be taken from the Class certification report and compliance verified during plan review.

(8)   Electrical arrangements
The next paragraph in MSC/Circ.1165, as amended by IMO MSC.1/Circs.1237 and 1269, concerns electrical equipment and arrangements. The electrical components of the pressure source for

the system are required to have a minimum rating of IP 54. In addition, the system should be supplied by both the main and emergency sources of power and should be provided with an automatic change-over switch. The emergency power supply should be provided from outside the protected machinery space.

(9)   Secondary water supply

The system should be provided with a redundant means of pumping or otherwise supplying the water-based extinguishing medium. The system should be fitted with a permanent sea inlet and should be capable of continuous operation using sea water in accordance with paragraph 17 of MSC/Circ.1165, as amended by IMO MSC.1/Circs.1237 and 1269.

(10)   System calculations

Paragraph 18 of MSC/Circ.1165, as amended by IMO MSC.1/Circs.1237 and 1269, requires that the system should be sized in accordance with a hydraulic calculation technique. Accordingly, the distribution system must be properly analysed to confirm that the required flow rates at the most hydraulically demanding system will be provided. System flow and pressure drop calculations are to be requested and reviewed (see above for a general discussion regarding this subject). In accordance with paragraph 19 of MSC/Circ.1165, as amended by IMO MSC.1/Circs.1237 and 1269, systems capable of supplying water at the full discharge rate for 30 minutes may be grouped into separate sections within a protected space. However, the sectioning of a system with such spaces must be specifically approved. In addition, paragraph 19 of MSC/Circ.668 indicates that regardless of the sectioning arrangements, the capacity and design of a system should be based upon the complete protection of the space demanding the greatest volume of water. Accordingly, even if sectioning of the system within a particular space is permitted, the system still needs to be capable of supplying all nozzles within that space are the required pressure. As indicated earlier, paragraph 20 of MSC/Circ.1165, as amended by IMO MSC.1/Circs.1237 and 1269, requires that the capacity and design of a system should be based upon the complete protection of the space demanding the greatest volume of water. Accordingly, the calculations should evaluate the most hydraulically demanding space, as well as the space demanding the greatest volume of water.

(11)   System component locations

In accordance with paragraph 22 of MSC/Circ.1165, as amended by IMO MSC.1/Circs.1237 and 1269, the pressure source components (e.g. pump and controls) must be located outside of the protected space. Accordingly, the arrangements for the pump and accumulator location, the power supply to the pump and controls, access and ventilation arrangements, etc., for the water mist system, must be examined to confirm compliance.

(12) Testing arrangements
A means for testing the operation and flow of the system to assure the required pressure and flow should be provided in accordance with paragraph 23 of MSC/Circ.1165, as amended by IMO MSC.1/ Circs.1237 and 1269.

(13) Audible and visual alarms
Activation of any water distribution valve should give a visual and audible alarm in the protected space and at a continuously manned central control station. An alarm in the central control station should indicate the specific valve activated paragraph 24 of MSC/ Circ.1165, as amended by IMO MSC.1/Circs.1237 and 1269.

(14) Operating instructions
Operating instructions for the system must be displayed at each operating position. The operating instructions are required to be in the official language of the Flag State, and if the language is neither English nor French, a translation into one of these languages should be included as per paragraph 25 of MSC/Circ.1165, as amended by IMO MSC.1/Circs.1237 and 1269.

(15) Spares
Spare parts and maintenance instructions for the system are to be provided as recommended by the manufacturer, in accordance with paragraph 26 of MSC/Circ.1165, as amended by IMO MSC.1/ Circs.1237 and 1269.

(16) Additives
Paragraph 27 of MSC/Circ.1165, as amended by IMO MSC.1/ Circs.1237 and 1269, requires that additives should not be used for the protection of normally occupied spaces unless they have been approved for fire protection service by an independent authority. The approval should consider possible adverse health effects to exposed personnel, including inhalation toxicity.

Fixed water mist systems in accommodation and service spaces

Like the earlier discussions in Chapter 5, the IMO developed Resolution A.800(19) as amended by MSC.265(84) and MSC.284(86), to provide guidelines for the approval of alternative water-based fire extinguishing systems as equivalent to the water sprinkler systems complying with FSS Code, Chapter 8. It is through compliance with the design and testing requirements specified in IMO Resolution A.800(19) as amended by MSC.265(84) and MSC.284(86) that water mist systems may be considered as fire protection systems, which can provide an equivalent level of safety and therefore an equivalent system to the water sprinkler systems discussed in the FSS Code, Chapter 8.

(1) General

Water mist systems are specialised systems. The characteristics of these systems (e.g., required minimum pressure, droplet size and velocity, and minimum required discharge flow rate) can vary substantially from manufacturer to manufacturer and there are numbers of different approaches used to generate water mist. The nozzles are particularly critical to the proper operation and effectiveness of a water mist system.

(2) Water mist nozzle and system testing and approval

Like the water mist component and system requirements discussed earlier for machinery spaces, Appendix '1' of IMO Resolution A.800(19), as amended by MSC.265(84) and MSC.284(86), provides substantial design and testing requirements specifically for the water mist nozzles. In addition, Appendix '2' identifies a series of fire tests that are required to qualify the nozzle (in particular) and the system (in general) for suitability to extinguish a number of different types of fires involving accommodation and service spaces. As previously discussed, fixed fire extinguishing system components are required to be Class certified and procedures similar to those discussed earlier should be followed for the certification of a water mist system and its components for use in accommodations and service spaces. The certification report should only be issued after verifying compliance with all design and testing requirements specified in appendixes '1' and '2' of IMO Resolution A.800(19) as amended by MSC.265(84) and MSC.284(86). The report should clearly indicate complete details regarding the basis of the system's approval including any details that could possibly affect the system's performance. As a minimum, these details should include:

(a) The system's design parameters (e.g., size and class of accommodation and service space for which it is approved, maximum enclosure height, models of nozzles, minimum nozzle pressures, nozzle flow rates particle size distribution, pumps and control valves used for the test, and maximum ventilation condition);

(b) Installation criteria used during the testing (e.g., nozzle spacing, minimum and maximum heights, and required stand-off); and

(c) Any other special issues or requirements which could potentially affect the satisfactory performance of future installations. The recording of this information is critical since compliance with such details must be verified during the review of a specific water mist system installation. The certification report should also list the various system drawings, schematics, and other documentation used for the approval (complete with reference numbers, dates, etc.) so that it can be determined if future submissions are utilising the same system design. Further discussions regarding the detailed testing and approval procedures of the individual manufacturer's system and nozzles in Appendices '1' and '2' are beyond the scope of this chapter. However, it is important for the engineer conducting the individual system review for a particular installation to understand that there are substantial design and testing requirements applicable to the water mist system and nozzles. The manufacturer's system must pass a series of fire tests applicable to the specific size accommodation and service space simply to qualify for consideration as specified in paragraph 3.2 of IMO Resolution A.800(19) as amended by MSC.265(84) and MSC.284(86). Review of a particular system installation must verify compliance with the information/criteria included on the Class issued certification report.

(3)   Automatic operation
The system should operate automatically, with no human action necessary to set it into operation, in accordance with paragraph 3.1 of IMO Resolution A.800(19) as amended by MSC.265(84) and MSC.284(86). Further, the system is required to be capable of both detecting the fire and acting to control or suppress the fire, in accordance with paragraph 3.2 of IMO Resolution A.800(19) as amended by MSC.265(84) and MSC.284(86). The detection and automatic operation are normally accomplished by the use of nozzles with heat-sensitive bulbs or links designed to allow the nozzles to open when the temperature in the vicinity of the nozzle reaches a certain activation temperature. The system is required to be capable of continuously supplying the extinguishing medium for a minimum of 30 minutes, and a pressure tank should be provided to meet the functional requirements specified in FSS Code, Chapter 8/2.3.2.1. On the basis of the above requirement, the system must be capable of automatic operation in the event of a fire, and the pump, pump controls, valving, etc. must be designed for such operation.

(4)   Accumulator tank
In association with the aforementioned requirement, the system should be fitted with a freshwater pressure tank to maintain the pressurised condition of the system. The tank should contain a standing charge of fresh water, equivalent to the amount of water, which would be discharged in one minute by the pump.

In addition, arrangements are to provide for maintaining an air pressure in the tank such that when the standing charge of fresh water in the tank has been used, the pressure will be not less than the working pressure of the sprinkler plus the pressure exerted by a head of water measured from the bottom of the tank to the highest sprinkler in the system. To accommodate the volume of air necessary to maintain the pressure throughout the one minute of discharge, the tank is required to have an actual volume equal to at least twice that of the charge of water specified earlier. In addition, suitable means of replenishing the air under pressure and of replenishing the freshwater charge in the tank are to be provided. Also, a glass gauge is required to be provided to indicate the correct level of the water in the tank. The accumulator tank addressed earlier should be considered as a pressure vessel and should comply with the applicable requirements set by Class. These requirements will vary depending upon the maximum pressure involved. During the review of the system, the applicable requirements set by Class should be identified and the submitter advised accordingly unless Class certification is already indicated on one of the drawing notes.

(5)   Wet pipe system
The system should be of a wet pipe design, but small exposed sections are allowed to be of the dry pipe, pre-action, deluge, antifreeze, or other type, where determined necessary. Refer to Section 2 of IMO Resolution A.800(19) as amended by MSC.265(84) and MSC.284(86) for appropriate definitions of these terms.

(6)   Design and suitability of components
In accordance with paragraph 3.6 of IMO Resolution A.800(19), as amended by MSC.265(84) and MSC.284(86), the system and its components are to be suitably designed to withstand ambient temperature changes, vibration, humidity, shock, impact, clogging, and corrosion normally encountered in vessels. Further, components within the protected spaces should be designed to withstand the elevated temperatures, which could occur during a fire. Paragraph 3.7 of IMO Resolution A.800(19), as amended by MSC.265(84) and MSC.284(86), requires that the system and its components 'be designed and installed in accordance with international standards acceptable to the Organisation and manufactured and tested to the satisfaction of the Administration in accordance with the appropriate elements of Appendices 1 and 2' In this regard, a footnote is also provided, which indicates that 'Pending the development of international standards acceptable to the Organisation, national standards as prescribed by the Administration should be applied'.

Like MSC/Circ.1165, as amended by IMO MSC.1/Circs.1237 and 1269, IMO Resolution A.800(19), as amended by MSC.265(84) and MSC.284(86), is focused on the design and testing requirements for the nozzle, as well as some overall system requirements. Accordingly, there are many areas of a water mist system which are not adequately addressed in the Resolution. To address such concerns, the Resolution references 'international standards acceptable to the Organisation'. However, currently there are no particular 'international standards acceptable to the Organisation', and therefore, the footnote allows for the consideration of appropriate national standards. While other appropriate national standards may exist, the national standard frequently used is NFPA 750, *Standard on water mist fire protection systems*. Accordingly, based on the aforementioned reference, the engineer conducting the review should understand that in addition to compliance with the design and testing requirements specified in Appendices '1' and '2' of IMO Resolution A.800(19), as amended by MSC.265(84) and MSC.284(86) for the system, as well as its components, the system and components must also comply with the requirements specified in an appropriate national or international standard, such as NFPA 750. The review should verify compliance with all aspects accordingly.

(7) Power supply arrangements
The system should be provided with power from both the main and emergency sources of power in accordance with paragraph 3.8 of IMO Resolution A.800(19) as amended by MSC.265(84) and MSC.284(86).

(8) Redundant pumping arrangements
The system should be provided with a redundant means of pumping or otherwise supplying the water-based extinguishing agent to the sprinkler system in accordance with paragraph 3.9 of IMO Resolution A.800(19) as amended by MSC.265(84) and MSC.284(86).

(9) Sea inlet
The system should be fitted with a permanent sea inlet and should be capable of continuous operation using sea water, in accordance with paragraph 3.10 of IMO Resolution A.800(19) as amended by MSC.265(84) and MSC.284(86).

(10) Calculations
Paragraph 3.11 of IMO Resolution A.800(19), as amended by MSC.265(84) and MSC.284(86), requires the system to be properly sized in accordance with a hydraulic calculation technique. This requires that the pressure sources and distribution system must be properly analysed to confirm that the required flow rates at the most hydraulically demanding system will be provided. System flow and

pressure drop calculations are to be requested and reviewed (refer to Chapter 5 for a general discussion regarding this subject).

(11) Zoning

The sprinklers are to be grouped into separate sections and no one section should serve more than two decks of one main vertical zone, as per paragraph 3.12 of IMO Resolution A.800(19) as amended by MSC.265(84) and MSC.284(86).

(12) Isolation valve

Each sprinkler section should be capable of being isolated by only one stop valve. The stop valve in each section should be readily accessible in a location outside of the associated section or in cabinets within stairway enclosures and its location clearly and permanently indicated, in accordance with paragraph 3.13 of IMO Resolution A.800(19) as amended by MSC.265(84) and MSC.284(86).

(13) Dedicated system

The sprinkler system piping is not to be used for any other purpose, in accordance with paragraph 3.14 of IMO Resolution A.800(19) as amended by MSC.265(84) and MSC.284(86). Accordingly, no connections for wash-down arrangements, deck showers, or other non-potable water services are to be permitted.

(14) System component locations

In accordance with paragraph 3.15 of IMO Resolution A.800(19), as amended by MSC.265(84) and MSC.284(86), the sprinkler system supply components (e.g., pump, valves, and controls) are to be located outside of any Category 'A' machinery spaces and not situated in any space required to be protected by the sprinkler system.

(15) Testing arrangements

Arrangements are to be provided, which will allow testing of the automatic operation of the system and confirm that the required pressure and flow will be provided, in accordance with paragraph 3.16 of IMO Resolution A.800(19) as amended by MSC.265(84) and MSC.284(86).

(16) Visual and audible alarms

In accordance with paragraph 3.17 of IMO Resolution A.800(19), as amended by MSC.265(84) and MSC.284(86), each sprinkler section should be provided with the means to give a visual and audible alarm at a continuously manned central control station within one minute of flow from one or more sprinklers. Each sprinkler section should also be fitted to a check valve, a pressure gauge, and a test connection with a means of drainage. An alarm in the central control station should indicate the specific valve activated, in accordance with paragraph 24 of MSC/Circ.1165, as amended by IMO MSC.1/Circs.1237 and 1269.

(17) Sprinkler control plan
A sprinkler control plan should be displayed at each centrally manned control station, in accordance with paragraph 3.18 of IMO Resolution A.800(19) as amended by MSC.265(84) and MSC.284(86).

(18) Installation plans and operating manuals
Installation plans and operating manuals for the system are required to be readily available onboard. A list or plan should be displayed showing the spaces covered and the locations of the zones in respect to each other. Instructions for testing and maintenance are to also be available onboard. See paragraph 3.19 of IMO Resolution A.800(19) as amended by MSC.265(84) and MSC.284(86).

(19) Nozzles
In addition to the previously discussed requirements for nozzles, the sprinklers are to have fast response characteristics as defined in ISO 6182–1. See paragraph 3.20 of IMO Resolution A.800(19) as amended by MSC.265(84) and MSC.284(86). Documentation verifying this should be submitted. Within accommodation and service spaces, paragraph 3.21 of IMO Resolution A.800(19), as amended by MSC.265(84) and MSC.284(86), requires that the sprinklers are to have a nominal temperature range of 57 °C (134 °F) to 79 °C (174 °F), except for in locations such as drying rooms, where high ambient temperatures might be expected, the operating temperature may be increased by not more than 30 °C (86 °F) above the maximum deck head temperature. The nozzle location, type of nozzle, and nozzle characteristics are to be within the limits tested during the Appendix '2' fire tests in accordance with paragraph 3.23 of IMO Resolution A.800(19) as amended by MSC.265(84) and MSC.284(86). This is part of the information which should be taken from the Class certification report.

(20) Total system capacity requirements
Like the requirements specified in the FSS Code, Chapter 8/2.3.3.2, paragraph 3.22 of IMO Resolution A.800(19) as amended by MSC.265(84) and MSC.284(86) requires that the pumps and alternative supply components are to be sized so as to be capable of maintaining the required flow to the hydraulically most demanding area of not less than 280 m^2 (3,012 ft^2). However, paragraph 3.22 permits the use of alternative minimum coverage areas if the total protected area is less than 280 m^2 (3,012 ft^2). The ability of the pump and the piping system to support this flow rate at the most hydraulically remote section must be verified through a review of the pump curves and appropriate pressure drop calculations as discussed earlier.

(21) Atrium
In accordance with paragraph 3.24 of IMO Resolution A.800(19), as amended by MSC.265(84) and MSC.284(86), for atrium with intermediate-level deck openings exceeding 100 m², ceiling-mounted sprinklers are not required.

(22) Level of protection
Paragraph 3.25 of IMO Resolution A.800(19), as amended by MSC.265(84) and MSC.284(86), requires the systems to be designed in such a way that during a fire occurrence, the level of protection provided to those spaces unaffected by fire is not reduced.

(23) Quantity of spare water mist nozzles
In accordance with paragraph 3.26 of IMO Resolution A.800(19) as amended by MSC.265(84) and MSC.284(86), a quantity of spare water mist nozzles should be carried for all types and ratings installed on the ship as follows:
< 300 total number of six spare nozzles
300 to 1,000 total number of 12 spare nozzles
> 1,000 total number of 24 spare nozzles.
The number of spare nozzles of any type need not exceed the total number of nozzles installed of that type.

(24) Protection against freezing
Paragraph 3.27 of IMO Resolution A.800(19), as amended by MSC.265(84) and MSC.284(86), requires that any parts of the system, which may be subjected to freezing temperatures in service, are to be suitably protected against freezing.

Fixed pressure water-spraying fire extinguishing system on cabin balconies of passenger ships

A fixed pressure water-spraying fire extinguishing system complying with the provisions of the FSS Code is required to be installed on cabin balconies of ships to which SOLAS regulation II-2/5.3.4 applies, where furniture and furnishings on such balconies are not as defined in regulations 3.40.1, 3.40.2, 3.40.3, 3.40.6, and 3.40.7 of SOLAS Chapter II-2.

Throughout this chapter we have discussed, at some length, the use of water as an extinguishing media for shipboard fires. Whilst water is a very effective firefighting media, it is not without its limitations, which must always be considered in any firefighting situation. One of the main alternatives to using water is foam, of which there are various forms. In the next chapter, we will examine the use of foam as a fire extinguishing media, the various types of foam, its advantages, and, of course, its limitations.

Chapter 6

Foam extinguishing systems

Foam is produced by the combination of three materials: water, air, and a foam-making agent. Foam is formed by first mixing the foam-making agent (i.e. foam concentrate) with water to create a foam solution. The actual foam bubbles are created by introducing air into the foam solution through an aerating device. The correctly chosen foam concentrate, when properly proportioned with water and expanded with air through an application device, will form a finished foam. The foam concentrate must be thoroughly mixed with water at a specific concentration ratio to produce the foam solution needed to create the desired foam consistency. Two of the most common concentrations used on ships are 3% and 6% foams. These values are the percentages of the concentrate to be used in making the foam solution. Thus, if 3% concentrate is used, three parts of concentrate must be mixed with 97 parts of water to make 100 parts of foam solution. If 6% concentrate is used, then six parts of concentrate must be mixed with 94 parts of water to make 100 parts of foam solution.

Once the desired foam solution has been mixed, the firefighting foam is then deployed to form a blanket over the surface of burning liquids. This blanket prevents flammable vapours from leaving the surface and prevents oxygen from reaching the fuel. As per Chapter 2, fire cannot exist when the fuel and oxygen are separated, and therefore, a properly placed foam blanket will smother the fire. In addition, the water in the foam has a cooling effect, which enables the foam to cool the surrounding structure to help prevent flash back. Ideally, the foam should flow freely enough to cover a surface rapidly yet have adequate cohesive properties to stick together sufficiently to establish and maintain a vapour tight blanket. In addition, the solution must retain sufficient water to provide a long-lasting seal. This is because a rapid loss of water will cause the foam to dry out and break down (wither) from the high temperatures associated with the fire. The foam should also be light enough to float on top of any flammable liquids yet be heavy enough to resist blowing away with wind.

DOI: 10.1201/9781003385523-8

TYPES OF FOAM AND FOAM CHARACTERISTICS

Foam must contain the right blend of physical characteristics to be effective. These characteristics include:

- *Knockdown speed and flow.* The ability of the foam blanket to spread across a fuel surface or around obstacles and wreckage, in order to achieve complete extinguishing effect, is critically important. Moreover, the foam must have good cohesion properties to maintain the blanket effect yet at the same time, must not be so viscous to hinder the ability of the foam to flow over the area, and form a self-supporting blanket.
- *Heat resistance.* The foam must be able to resist the destructive effects of heat radiated from any remaining fire from the liquid's flammable vapour and any hot metal wreckage or other affected objects in the area.
- *Fuel resistance.* An effective foam will minimise fuel pick-up so that the foam does not become saturated and burn off.
- *Vapour suppression.* A vapour-tight blanket is a critical function of a foam's ability to extinguish a fire. The foam produced must be capable of suppressing the flammable vapours to break the fuel-oxygen-heat fire triangle and to minimise the risk of re-ignition.
- *Alcohol resistance.* Due to alcohol's affinity to water, and due to the foam blanket being between 97% and 94% water, foam blankets that are not alcohol-resistant will be destroyed if used on alcohol-based cargoes.

There are two basic types of foam: chemical and mechanical.

Chemical foams

Chemical foam is formed by mixing a solution of alkali (usually sodium bicarbonate), an acid (usually aluminium sulphate), water, and a stabiliser. The stabiliser is added to make the foam tenacious and increases its longevity. When these chemicals react, they form a foam or froth of bubbles filled with carbon dioxide gas. The carbon dioxide in the bubbles has little or no extinguishing value. Instead, its purpose is to inflate the bubbles. Between 7 and 16 volumes of foam are produced for each volume of water. Premixed foam powders may also be stored in cans and introduced into the water during firefighting operations. For this, a device called a *foam hopper* is typically used. Alternatively, the two chemicals may be premixed with water to form an aluminium sulphate solution and a sodium bicarbonate solution. The solutions are then stored in separate tanks until the foam is needed. At that time, the solutions are mixed to form the foam. Today, few chemical foam systems are still used in shipboard settings.

Mechanical (air) foams

Mechanical foam is produced by mixing a foam concentrate with water to produce a foam solution. The bubbles are formed by the turbulent mixing of air and the foam solution. As the name 'air foam' implies, the bubbles are filled with air. Aside from the workmanship and efficiency of the equipment, the degree of mixing determines the quality of the foam, whereas the design of the equipment determines the quantity of foam produced. There are several types of mechanical foams, similar in nature, with each having their own special firefighting capabilities. The main types of mechanical foams include the following:

Protein foam

Protein foams were the first type of mechanical foam to be manufactured and have been in continuous use since World War I. These foams are usually produced by the hydrolysis of waste protein material, such as protein-rich animal waste (i.e. hoofs and horns) and vegetable waste that is hydrolysed (i.e. subjected to a chemical reaction with water that produces a weak acid). In addition, stabilising additives and inhibitors, such as mineral salts, are added to prevent corrosion, resist bacterial decomposition, control viscosity, and increase the foam's resistance to withering from high temperatures. At the time of a fire, the protein foam concentrates are mixed with freshwater or seawater in 3% or 6% solutions. The foam concentrate can produce foam with all types of water, except that which is contaminated with oil. When antifreeze is added, foam can be produced in sub-zero temperatures (as low as –23.3 °C (–10 °F)).

Fluoroprotein (FP) foam

Fluoroprotein (FP) foams are formed by the addition of special fluorochemical surfactants with protein foam (fluorinated compound bonded to the protein). This enhances the properties of protein foam by increasing foam fluidity (ease of flow) and improves the properties of regular protein foam by providing faster knockdown and excellent fuel tolerance. FP foam works well with dry chemical agents, and when the water is mixed with antifreeze, produces foam in sub-zero temperatures.

Film forming fluoroprotein (FFFP) foam

Film forming fluoroprotein (FFFP) foams are a combination of fluorochemical surfactants with protein foam. They are designed to combine the burn back resistance of an FP foam with an increased knockdown power. These foams also release a film on the surface of the hydrocarbon, which prevents the vapours from igniting.

Aqueous film forming foam (AFFF)

Aqueous film forming foams (AFFFs) are a combination of fluorochemical surfactants and synthetic foaming agents that create a unique characteristic – an aqueous film. This film is a thin layer of foam solution with unique surface energy characteristics that spreads rapidly across the surface of a hydrocarbon fuel causing rapid fire knockdown. The aqueous film is produced by the action of the fluorochemical surfactant reducing the surface tension of the foam solution to a point where the solution is supported on the surface of the hydrocarbon. AFFFs are more effective on hydrocarbon (fuel)-related fires with higher surface tensions, such as kerosene, diesel oil, and jet fuels, but are less effective on fuels with lower surface tensions, such as hexane and high-octane petrol and gasoline. AFFFs may drain the foam solution (the water and foam concentrate mixture) rapidly from the foam bubble to produce optimum film formation for rapid fire extinguishing effect. To achieve these qualities, however, long-term seal and burn-back resistance are unfortunately sacrificed.

Alcohol resistant aqueous film-forming foam (AR-AFFF)

Alcohol resistant-aqueous film forming foams (AFFFs) are produced from a combination of synthetic stabilisers, foaming agents, fluorochemicals, and alcohol-resistant membrane-forming additives. Polar solvents and water miscible fuels, such as alcohols, are destructive to non-alcohol resistant-type foams. Alcohol aggressively mixes with the water in the foam destroying the foam blanket and its firefighting properties. Instead, alcohol-resistant foams act as a conventional AFFF on hydrocarbon fuels, forming an aqueous film on the surface of the hydrocarbon. When used on alcohol-type fuels, the membrane-forming additives form a tough polymeric (sometimes called *mucoloid*) membrane which separates the foam from the alcohol and prevents the destruction of the foam blanket. While some concentrates are designed for use on alcohol-type fuels at 6% and hydrocarbon fuels at 3%, modern formulations are designed to be used with 3% on both fuel groups. These newer formulations provide more cost-effective protection against alcohol-type fuels, using half the amount of concentrate as a 6% agent. The use of a 3 × 3 AR-AFFF also simplifies setting the proportioning percentage at an incident, since it is always 3%. Overall, AR-AFFFs are the most versatile type of foam available today, offering good burn back resistance, knockdown, and high fuel tolerance on both hydrocarbon- and alcohol-related fuel fires.

Synthetic detergent-based foam

Synthetic detergent-based foam consists of alkyl sulphates. This foam has less burn back resistance than protein formulas but may be used with all types of dry chemicals. It foams more readily than is the case with natural proteins and requires less water. This is important where water supplies are limited.

Foam expansion ratios

There are three types of expansion foam ratios, depending on the speed and surface area of each foam type.

Low expansion foams

Low expansion foams are defined as those foams with an expansion ratio of 12:1 when mixed with air. That is, one volume of foam concentrate will create 12 volumes of foam. Low expansion foams are effective in controlling and extinguishing most flammable liquid (Class 'B') fires. Foams typically used on tanker deck foam systems are of the low expansion foam type.

Mid expansion foams

Mid expansion foams refer to those foams with an expansion ratio of between 20:1 and 100:1. Few applications of mid expansion foams are found in shipboard applications.

High expansion foams

High expansion foams are those which expand in ratios over 100:1. Most systems produce expansion ratios of 400:1 to 1,000:1. Unlike conventional foam, which provides a blanket a few centimetres over the burning surface, high expansion foam is three dimensional, which means it is measured in terms of its length, width, height, and cubic area. High expansion foam is especially useful for firefighting in confined spaces. As it is heavier than air, but lighter than oil or water, it will flow down openings and fill compartments, spaces, and crevices, therein replacing the air in these spaces. Doing so, the foam deprives the fire of crucial oxygen causing it to suffocate. Because of its water content, the foam absorbs heat from the fire and cools the burning material. When the high expansion foam has absorbed sufficient heat to turn the water content in the foam to steam (100 °C (212 °F), the steam continues to replace the oxygen, thus combating the fire.

LIMITATIONS AND ADVANTAGES OF FOAM AS AN EXTINGUISHING AGENT

As with all firefighting systems, there are both limitations and advantages to using foam as an extinguishing agent. When deployed properly, foams offer an effective method of extinguishing fires; however, it is also important to remember that, given foams are aqueous (water-based) in nature, they are electrically conductive and should not be used on live electrical equipment or electrical type fires. Like water, foams should never be used on combustible metal fires as the water can cause second-ary reactions. Many types of foam must not be used with dry chemical extinguishing agents. Only AFFF is the exception to this rule as it may be used in a joint attack with dry chemical powder. Foams are not suit-able for fires involving gases and cryogenic (extremely low temperature) liquids. If foam is placed on burning liquids (such as asphalts) whose temperatures exceed 100 °C (212 °F), the water content of the foam may cause frothing, spattering, or slop-over. Slop-over is different from boil-over, although the terms are frequently confused. Boil-over occurs when the heat from a fire within a tank travels down to the bottom of the tank causing the water that is there to boil and push part of the tank's contents over the side. Certain oils with a high water content, such as crude oil, have a notorious reputation for boil-over. Slop-over occurs when foam, introduced into a tank of hot oil [where the surface temperature is over 100 °C (212 °F)], sheds its water content due to the high temperature. The water forms an emulsion of steam, air, and foam. The forming of the emulsion is accompanied by a corresponding increase in volume. Since tanks are three dimensional, the only place for the emulsion to go is over the sides of open tanks or into the vents of enclosed tanks. This can cause considerable, and expensive, secondary damage. Last of all, there must be a sufficient volume of foam available to cover the entire surface of the burning material. In addition, there should be enough foam to replace foam that is burned off and to seal breaks in the foam surface. Not doing so will inevitably result in a failure to extinguish the fire properly.

Despite the not inconsiderable limitations associated with foam, foam is quite effective in combating Class 'B' and some Class 'A' fires. This is because foam is a very effective smothering agent and provides cooling as a second-ary effect. Foam creates a vapour barrier, which prevents flammable vapours from rising from the liquid surface. The surface of an exposed tank can be covered with foam to protect it from a fire in neighbouring tank(s). Foam is of some use on Class 'A' fires because of its water content; AFFF is especially effective, as are certain types of wet-water foam. Wet-water foam is made from detergents, and its water content quickly runs out and seeps into the burning material. It is not usually found in shipboard settings but is com-mon on shoreside locations such as bulk storage warehouses and on piers and quays. For oil-related incidents, foam is effective in blanketing oil spills. However, if the oil is running, attempts should be made to shut down the affected valves, where such action will stop the flow of oil. If that is not pos-sible, the flow of oil should be dammed instead. Foam should be applied on the upstream side of the dam (to extinguish the fire) and on the downstream side (to place a protective cover over any oil that has seeped through). Last of all, foam is the most effective extinguishing agent for fires involving large tanks of flammable liquids.

Some other ancillary benefits of using foam include the fact that it can be made with both freshwater or seawater, and hard or soft water. Furthermore, foam does not break down readily, and it extinguishes fire progressively when applied at a consistent rate. Foam stays in place, covers, and absorbs heat from materials that could potentially cause re-ignition. Last of all, foam concentrates are not heavy, and their systems are considerably more confined when compared to water-based extinguishing systems.

BASIC GUIDELINES FOR THE USE OF FOAM AS AN EXTINGUISHING AGENT

For foam to provide maximum effect during firefighting, it is imperative that it is stored according to the manufacturer's instructions and deployed accordingly. To ensure the foam solution is kept in prime condition, it is important for these basic guidelines to be followed.

(1) Storage

If manufacturer recommendations are followed, then protein or synthetic foam concentrates should be ready for active service even after many years of storage.

(2) Water temperature and contaminants

Foams in general are more stable when generated with lower temperature water. Although all foam liquids will work with water in excess of 37.7 °C (100 °F), the typical concentrate works best with water in the temperature range between 1.7 °C and 26.7 °C (35 °F and 80 °F). Either fresh or sea water may be used. Water containing known foam contaminants, such as detergents, oil residues, or certain corrosion inhibitors, may adversely affect foam quality.

(3) Combustible products in air

It is desirable to take clean air into the foam nozzle at all times, although the effect of contaminated air on foam quality is minor with low expansion foams.

(4) Water pressures

Nozzle pressures should be held between 3.4 bar and 13.8 bar (50 and 200 psi). If a proportioner is used, proportioner pressure should not exceed 13.8 bar (200 psi). Foam quality deteriorates at higher pressures. Range falls off at lower pressures.

(5) Non-ignited spills

Where flammable liquids have spilled, fires can be prevented by prompt coverage of the spill with a foam blanket. Additional foam may be necessary from time to time, to maintain the blanket for extended periods until the spill has been cleaned up.

(6)   Electrical fires
      Foam should be considered nearly the same as water when used on electrical fires and is therefore not generally recommended for use on electrical fires. However, if the supply of current to the electrical circuits can be interrupted or broken, then foam can be used to extinguish such fires.

(7)   Vapourised liquids
      Foam is not recommended for use on materials that may be stored as liquids, but are normally vapour at ambient conditions, such as propane, butadiene, and vinyl chloride. Firefighting foam is not recommended for use on materials that react with water, such as magnesium, titanium, potassium, lithium, calcium, zirconium, sodium, and zinc.

FOAM SYSTEM EQUIPMENT

As mentioned earlier, finished foam is a combination of foam concentrate, water, and air. When these components are brought together in proper proportions and thoroughly mixed, foam is produced. To ensure the foam solution is properly mixed and defined, the correct equipment must be used.

Proportioning devices

All foam proportioners are designed to introduce the proper percentage of foam concentrate into the water stream. There are several varieties of proportioning systems available to the fire service today. The choices range from the more commonly used and economical in-line eductors to *around-the-pump systems* to the sophisticated and more expensive *balanced pressure systems*.

Eductors

Eductors are the most common form of proportioning equipment. They are used for dedicated foam discharges and around-the-pump systems. Eductors work on the Venturi principle. Water is introduced, under pressure, at the inlet of the eductor. The eductor reduces the orifice available for the water to pass through, so it must speed up to get through. This creates a pressure drop that, in turn, puts suction on the pick-up tube. As the foam concentrate is pulled up the tube, it passes through a metering valve that allows the correct percentage to be introduced into the water stream. In most cases, the metering valve can be adjusted to select a 1, 3, or 6% foam solution. Eductors are extremely reliable and simple pieces of equipment. However, they do have certain limitations.

- *Eductor GPM flow rate restrictions.* All eductors have litres per minute (gallons per minute (gpm) solution flow rating. Typically, 227, 360, 473, 946, l/min (60, 95, 125, 250 gpm) models are available. The eductor must be matched with a nozzle that has the same flow rating. Eductor/nozzle mismatches are the most common cause of fire service proportioning problems. Mismatches can result in a weak solution or a complete shut-down of foam concentrate pick-up.
- *Inlet pressure requirements.* Eductors establish their pressure drop at a fairly high energy cost. The loss between the inlet and outlet pressure of an eductor can be 40% or more. In order to accommodate this loss and still provide adequate nozzle pressure, relatively high eductor inlet pressures are necessary. Most manufacturers recommend inlet pressures at the eductor in the range of 12.4–13.8 bar (180–200 psi).
- *Most eductors will continue to pick up at lower inlet pressures.* However, at these lower pressures, the solution flow drops. Under these conditions, it becomes impossible to accurately know the concentration of the foam solution being delivered to the fire.
- *Back pressure restrictions.* Too much back pressure on an eductor can shut down foam concentrate pickup. Therefore:
  o The nozzle and eductor must be matched.
  o The nozzle must be fully opened or fully closed. It cannot be in between.
  o Prevent kinks in the hose line between nozzle and eductor.
  o The nozzle should not be elevated above the eductor.
  o The hose lay cannot exceed manufacturer's recommendation.
  Following these simple rules helps to eliminate excessive back pressure on the eductor.
- *Eductors must be kept clean.* Eductors must be thoroughly cleaned after each use. Failure to clean an eductor can result in clogging and blockage due to hardening foam concentrate residue. If this occurs, the eductor will not function properly, if at all. When eductors are *properly* understood and maintained, they can accurately and reliably proportion foam at relatively low cost.

Around-the-pump systems

Another method of proportioning is the around-the-pump-type system. In this case, an eductor is installed on the discharge side of the water pump. As before, water flow causes a vacuum that picks up and introduces the foam concentrate into the pump suction. An adjustable metering valve controls the flow of the foam concentrate. Around-the-pump systems offer several advantages when compared to an in-line eductor:

- Variable Flow Rate. The discharge rate can be adjusted for the specific application. The rate is infinitely variable up to the maximum flow of the unit.
- Variable Pressure. The system operates at any pressure above 8.6 bar (125 psi). The pump operation is the same with foam or water.
- No Back Pressure Restrictions. The unit is not affected by hose length or elevation loss.
- No Nozzle Restrictions. The unit operates with any size or type of nozzle.

However, Around-the-Pump systems have their own limitations:

- Pump inlet pressure is limited to 0.7 bar (10 psi) to prevent a back pressure condition that will shut the system down.
- There is no choice of simultaneous flow of foam solution and plain water.
- An operator must continually calculate, set, and monitor the foam proportioning metering valve to correspond with the volume [l/min (gpm)] being flowed.
- Clean-up time can be long since ALL discharges must be flushed, whether or not they were opened during the operation.

Balanced pressure foam proportioners

Balanced pressure systems are extremely versatile and accurate. Most often, these systems are associated with fixed systems and specialised mobile equipment. Their design and operation are complex. The principle of operation is based on the use of a modified venturi proportioner commonly called a *ratio controller*. As water passes through a jet at the inlet of the ratio controller, it creates a reduced pressure area between the jet and a downstream section called a throat or receiver. This reduction in pressure causes foam concentrate to flow through a metering orifice and into the reduced pressure area. As the water flow through the ratio controller jet increases, so does the level of pressure reduction, thereby affecting a corresponding pressure drop across the foam liquid metering orifice. This corresponding pressure drop results in a foam liquid flow, which is proportionate to the water flow through the ratio controller. As both the water and foam liquid flow into a common reduced pressure area, it is necessary only to maintain identical water and foam liquid pressures at the inlets of the ratio controller. Pressure sensing lines lead from the foam liquid and water lines upstream of the ratio controller water and foam inlets to the diaphragm valve. This valve automatically adjusts the foam liquid pressure to correspond to the water pressure. A duplex gauge monitors balancing of foam liquid and water pressures on a single gauge. For manual operation, the diaphragm valve is not required. The pressure of the foam liquid is adjusted to correspond to the water pressure by means of a manually operated valve in the foam liquid bypass piping. The pressure loss across the proportioner is approximately 1.7–2.1 bar (25–30 psi) at maximum flow, depending on the ratio controller size selected. The minimum flow for which this device will proportion correctly is approximately 15% of the maximum flow for which it is designed. Balanced proportioning allows for a wide range of flows and pressures without manual adjustments, while placing no limitations on inlet pressure during foam operation.

Foam nozzles

For the most effective and economical use, the foam solution must be properly expanded. Standard fog nozzles generally do not provide optimum expansion, and therefore, do not provide for the best, most cost-effective application of the foam supply. In the case of polar solvent fuels, these standard fog nozzles may not deliver a foam quality that is able to extinguish the fire. Foam nozzles are specifically designed to air aspirate (expand) the foam solution and form finished foam. There are three main types of foam nozzles.

Low expansion

Low expansion nozzles expand foam solution up to 12:1 (i.e. for every 3.8 l (1 gal) of foam solution that enters the base of the nozzle, approximately 45.6 l (12 gal) of finished foam is produced). These nozzles draw air at the base of the nozzle, the air and the solution mix travel up the foam tube (this is called residence time), and the properly expanded foam exits the nozzle.

Medium expansion

Medium expansion nozzles can have expansion characteristics as high as 100:1, although expansions of 50:1 are more common. They operate in much the same way as low expansion nozzles. However, the diameter of the nozzle is much larger.

High expansion

High expansion foam nozzles can expand foam in excess of 100:1, when high expansion foam concentrates are used.

Foam monitors

The foam monitors or turrets are permanently installed foam discharge units capable of being aimed and projecting large quantities of foam substantial distances. They normally are mounted on a rotating base that allows the projection of foam in a 360-degree circle around the monitor platform. The angle of throw from the horizon can also be adjusted to facilitate flexibility in directing the foam to the fire. The foam solution is supplied to the monitor through a hard-piped foam main system that incorporates an expansion nozzle to aspirate the foam.

Applicators

Foam applicators are portable foam discharge devices supplied with foam solution through a hose from the hard-piped foam main. The applicators provide the flexibility to apply foam directly to specific locations or in a manner that the monitors may not be effective.

Valves and piping

The foam solution is distributed from the proportioning device to the monitors or applicators through a system of pipes and valves. The piping system must be adequately designed to match the flow rates of the equipment, and a thorough understanding of the system control valves is critical for quick and effective operation of the system. A diagram of the piping system and control valves is typically posted in the foam supply room and identifies which valves are to be opened in the event the system must be activated. The diagram normally explains thoroughly and clearly all the steps necessary to put the system into operation. Colour coding of the valves is also frequently used and aids in identification (e.g. all valves that are to be opened when a fire alarm is activated may be painted some distinctive colour). Each valve is also normally labelled with respect to its function, which assists in operating, restoring and maintaining the system.

Foam concentrate storage

The foam concentrate is stored in tanks ready to supply the proportioning system. The concentrate tank should be kept filled with liquid halfway into the expansion dome for prolonged storage life of the concentrate. The tank should be kept closed to the atmosphere, except for the pressure-vacuum vent. When a tank is partially empty, there is a larger liquid surface area to interact with air. This allows excessive evaporation and condensation, which degrade the foam concentrate and permit corrosion of the tank shell.

FOAM FIREFIGHTING APPLICATION TECHNIQUES

Bounce-off technique

When foam nozzles are used, particular care should be taken to apply the foam as gently as possible. For straight stream use, the foam should be bounced or banked off a wall or other obstruction when available.

Bank-in technique

Foam can also be rolled onto the fuel surface by hitting the ground in front of the spill and allowing the foam to 'pile up' in front of the spill. The velocity of the stream will roll the foam onto the fuel.

Rain-down technique

The foam nozzle is directed almost straight up, and the foam stream is allowed to reach its maximum height and break into small droplets. The nozzle operator must adjust the altitude of the nozzle, so the fallout pattern matches that of the spill area. This technique can provide a very fast and effective knockdown. However, if the fuel has had a significant pre-burn and a thermal column has developed, or if the weather is severe (high winds), the rain-down method may not be practical or effective.

Never plunge

Plunging the stream directly into the fire can splash the fuel causing the fire to spread. If a foam blanket exists, plunging can break the existing blanket allowing vapours to escape. This usually results in spreading the fire, reignition, or flare ups. Usually, the fire will lessen in intensity or self-extinguish once the plunging stream is removed. If the foam nozzle is equipped with a spray stream attachment, it should be used to provide the gentlest application possible and reduce the mixing of foam and fuel. Only as a last resort should a straight stream be directed into the centre of a pool or spill. Under this condition, the efficiency of the foam will be one-third or less than when applied by the recommended methods. Conventional AFFFs may be used effectively with standard water spray nozzles under some conditions although a very unstable foam with relatively poor reignition resistance is formed from such devices. Do not use water streams in such a way as to physically disrupt a foam blanket. Water streams may be used for cooling adjacent areas or as a fine spray to reduce flame radiant heat. However, do not direct water streams where a foam blanket has been or is being applied. The following information is provided as general guidance regarding Class requirements for foam fire extinguishing systems. However, reference should always be made to the Class rules applicable to the specific vessel concerned for the complete set of requirements.

GENERAL AND ADDITIONAL REQUIREMENTS FOR FOAM EXTINGUISHING SYSTEMS

General requirements applicable to all foam systems

All valves, fittings, and piping must comply with the applicable Class requirements. Accordingly, all valves and fittings should be designed and manufactured in accordance with a recognised standard, be suitable for the intended pressures, and comply with all other requirements set by Class. These components are also to be certified, as applicable, by Class.

Pipe and pipe joints

The design of the pipe and pipe joints must comply with Class requirements. Accordingly, compliance with the wall thickness requirements for the piping, as well as compliance of the type and design of the pipe joints, should be verified during plan review.

Materials

In addition to the common requirements and limitations, the materials used in the system should not be rendered ineffective by heat. This requirement is important so that the foam system will remain intact and functional even if a portion of the foam system piping is within the immediate vicinity of the fire. To be considered not 'readily rendered ineffective by heat', a component should be certified as having passed an acceptable recognised fire test or the material should have a melting temperature higher than the test temperature specified in an acceptable fire test. Accordingly, the review should verify that the materials of the valves, fittings, and pipe, as well as the method of joining sections of pipe, will not be rendered ineffective by heat.

Pumps

The pumps associated with the foam system are obviously critical to the satisfactory operation of the system, and therefore should be tested in the presence of a Class Surveyor.

Pressure vessels

In some designs, the concentrate storage tank utilises a bladder and is subjected to pump pressure to facilitate the discharge of the concentrate in water stream. Where such designs are used, the tank should be considered a pressure vessel and must comply with Class requirements, as applicable.

System component certification

Fixed fire extinguishing system components are to be Class certified. Accordingly, components such as foam system eductors, proportioners, monitors, and nozzles must comply with the certification requirements specified by Class, with the details verified accordingly.

Additional requirements for oil tanker deck foam systems

Separate provisions typically apply to oil tankers of 20,000 GRT and more. These include provisions whereby the cargo tank deck area must be fitted with appropriate protective measures and the cargo tanks with dedicated deck foam systems. For oil tankers less than 20,000 GRT, they must be fitted with a deck foam system or equivalent. Accordingly, almost all oil tanker carriers are required to be fitted with a deck foam system in accordance with the provisions of SOLAS regulation II-2/1.6.3, except where:

> *Liquid cargoes with a flashpoint exceeding 60°C [140°F] other than oil products or liquid cargoes subject to the requirements of the International Bulk Chemical Code are considered to constitute a low fire risk, not requiring the protection of a fixed foam extinguishing system.*

The deck foam system must be capable of discharging foam at the required rate to the entire cargo tank deck area, as well as into any cargo tank that is open due to a rupture in the deck. This system will normally consist of fixed monitors, as well as required hand line applicators.

Controls

The deck foam system must be capable of simple and rapid operation. This requires that the controls and arrangements for the system be designed so that it can be brought into operation quickly and would preclude complicated interconnections with non-firefighting systems. It is a requirement that the main control station for the system is located outside the cargo area, adjacent to the accommodation spaces, and is readily accessible and operable in the event of fire in the area being protected by the foam system. This is considered to require the control arrangements (controls for the motors, valves, pumps, etc. required to operate the system) to be in the location described earlier. The deck foam system concentrate tank, proportioner unit, etc. should also normally be installed in the above location but are not specifically mandated to be located at the control station and therefore may be in the engine room. The pump(s) supplying water are normally located in the engine room, but the controls for the pumps, as well as all portions of the foam system not located within the space containing the foam system control station, should be operable from the foam system control station.

Related IACS Unified Interpretation.
UI SC150 FSS Code, Chapter 14, 2.1.2 and 2.3.1.

The major equipment such as the foam concentrate tank and the pumps may be situated in the engine room. The controls of the system are to be situated in accordance with the FSS Code, Chapter 14, 2.3.1.

Application rate and system flow calculations

The minimum flow rate for the deck foam system for a typical oil carrier is determined by the criteria identified by Class requirement. Where so stated, these requirements should identify three different sets of criteria, and the minimum capacity of the system is not to be less than the largest of the coverage rates identified. Typically, these different sets of criteria are:

- The requirement for a system flow rate of not less than 0.6 l/min/m^2 (0.015 gpm/ft^2) based on the total cargo tank deck area. The cargo tank deck area is further defined as the maximum breadth of the vessel multiplied by the total longitudinal extent of the cargo tank spaces.
- The requirement for a system flow rate of not less than 6.0 l/min/m^2 (0.15 gpm/ft^2) based on the horizontal deck area of the single largest cargo tank.
- The requirement for a system flow rate of not less than 3.0 l/min/m^2 (0.075 gpm/ft^2) of the area protected by the largest monitor, but not less than 1,250 l/min.

As indicated earlier, the minimum flow rate of the system should not be less than the largest rate identified in the three sets of criteria specified in the Class requirements. Complete detailed system flow and pressure drop calculations are required to verify that the capacity and sizing of the system pumps, piping, and monitors are adequate to comply with the above requirements. There are certain cargoes on which regular foams are not effective and certain alternative requirements, including different application rates, are applicable. For more information, refer to the section on *Suitability of foam concentrate*.

Foam concentrate quantity

For a vessel fitted with an inert gas system, the amount of foam concentrate carried onboard must be sufficient to supply the system for a period of at least 20 minutes when operating at the system's maximum flow rate. For a vessel not fitted with an inert gas system, the amount of foam concentrate carried onboard must be sufficient to supply the system for a period of at least 30 minutes when operating at the system's maximum flow rate. The above is based on the average requirements stipulated in the FSS Code. The amount of foam concentrate required to be carried should be calculated based upon the maximum rated capacity of the system for the period specified earlier, the expansion ratio of the foam should not typically be greater that 12:1. However, if it is slightly greater than 12:1, the calculated amount is still to be based upon the 12:1 ratio. If a medium expansion foam (between 50:1 and 150:1) should be used, the same would require special Class approval.

System arrangements

On the basis of Class requirements, the cargo deck foam system should be a fixed system. Also, on tankers of 4,000 DWT and above, the system is to consist of monitors, as well as foam applicators. Tankers of less than 4,000 DWT may utilise a fixed system supplying only foam applicators with a capacity of not less than 25% of the foam supply rate or as stated in the section *Application rate/system flow calculations* earlier.

Foam monitors

The cargo deck foam monitors (see the general discussion earlier for definitions and descriptions) are critical components of the deck foam system as they provide the crew with the ability to engage the fire by placing large quantities of foam where needed from a relatively safe location. Class will generally provide specific requirements concerning the number, capacity, and locations of the monitors. Details of the monitor(s), as well as the proposed locations, should be checked regularly. As these requirements are generally shared by all Class administrations, these requirements are summarised here. Each monitor must have the capacity to deliver no less than 50% of the foam required by the Class rules. In addition, each monitor should have a capacity of at least 3.0 l/min/m² (0.075 gpm/ft²) of the area protected by that monitor, but in no case less than 1,250 l/min (330 gpm). The capacity of the monitors should be verified by reviewing the monitor capacity curves produced by the manufacturer at the pressure indicated in the foam piping system pressure drop calculations. Accordingly, pressure drop calculations for the foam piping distribution system should be requested and reviewed. The number and position of the monitors should be adequate to provide coverage to the entire cargo tank deck area, as well as into any cargo tank that is open due to a rupture in the deck. The coverage area of any specific monitor should be based upon the following:

(1) Only the area forward of the monitor should be considered protected by that specific monitor.
(2) The distance to the farthest point of the coverage of a monitor should not be more than 75% of the monitor through in still air.
(3) Monitors are to be provided on both the port and starboard sides of the vessel at the front of the accommodations space. Compliance regarding the locations and capacities of the monitors with the above requirements should be verified by reviewing a deck foam system arrangement drawing in conjunction with the monitor capacity and throw data provided by the monitor manufacturer along with system pressure flow drop calculations determining the pressure at the inlet of the monitor.

Foam applicators

The foam applicators (see general discussion earlier for definitions and descriptions) are also a critical component of the deck foam system as they provide the crew with the versatility and flexibility needed to combat a cargo tank fire. Class will provide specific requirements concerning the number, capacity, and locations of the applicators. Details of the applicators, as well as the proposed locations, should be regularly checked. As these requirements are generally shared by all Class administrations, these requirements are summarised here. Each applicator must have the capacity to deliver no less than 400 l/min (106 gpm) and a throw of at least 15 m (50 ft) in still air. Once again, the capacity and throw of the applicators should be verified by reviewing the capacity and throw curves produced by the manufacturer at the pressure indicated in the foam piping system pressure drop calculations. Each monitor station should be provided with a hose connection for an applicator, including the monitor stations provided at the port and starboard sides of the vessel at the front of the accommodations space. The number and locations of the foam main outlets for hose connections to serve the foam applicators should be sufficient so that at least two applicators can be directed to any part of the cargo tank deck area. At least four applicators with the appropriate hoses, connection devices, etc., must be carried onboard.

Isolation valves

The foam main should be fitted with isolation valves immediately forward of any monitor position so to provide the means to isolate damaged sections of the main. These isolation valves should be of a type, design, and material such that they will not be rendered ineffective by heat.

Simultaneous operation

The operation of the deck foam system at its required output should permit the simultaneous use of the minimum required number of jets of water at the required pressure from the fire main. This is especially important to note when the fire pumps are being utilised to also supply the foam system. If the fire pumps are used for this service, the combined capacity of the fire pumps must be capable of supplying the total required fire main capacity at the required pressures while simultaneously supplying the maximum capacity required by the foam system.

Related IACS Unified Interpretation:
UI SC61 re. FSS Code, Ch. 14, 2.1.3.

A common line for fire main and deck foam line can only be accepted provided it can be demonstrated that the hose nozzles can be effectively controlled by one person when supplied from the common line at a pressure needed for operation of the monitors. Additional foam concentrate should be provided for operation of two hose nozzles for the same period required for the foam system. The simultaneous use of the minimum required jets of water should be possible on deck over the full length of the vessel, in the accommodation, service spaces, control stations, and machinery spaces (MSC/Circ.1120).

Foam concentrate, quality, and testing

As discussed earlier, the quality of the foam and its suitability for use in the intended application is critical. Accordingly, the IMO has developed performance testing criteria for foam concentrates. This testing criterion is provided in IMO MSC.1/Circ.1312, 'Revised guidelines for the performance and testing criteria, and surveys of foam concentrates for fixed fire extinguishing systems'. The testing procedure covers a wide spectrum of requirements, from freezing and drain time testing to expansion ratio and fire testing. It is important to note that certain portions of the testing, such as the expansion ratio, drain time and fire tests, are dependent upon the types of mixing and expansion equipment used (proportioners, monitors, etc.). Accordingly, it should be verified that the foam concentrate has been tested in accordance with IMO MSC.1/Circ.1312 and further that the equipment being provided onboard is the same equipment utilised during the testing or is the equipment that will produce foam which is equivalent to that tested.

Suitability of foam concentrate

In addition to the aforementioned, it is critical to recognise that the foam to be used must be suitable for the specific cargoes to be carried.
   Related IACS Unified Interpretation:
   SOLAS (Consolidated Edition 2014) Regulation II-2/1.6.2.1 states

> *a liquid cargo with a flashpoint of less than 60°C for which a regular foam firefighting system complying with the FSS Code is not effective, is considered to be a cargo introducing additional fire hazards in this context. The following additional measures are required: the foam shall be of alcohol-resistant type. The type of foam concentrates for use in chemical tankers shall be to the satisfaction of the Administration, taking into account guidelines developed by the Organisation (Refer to IMO MSC.1/Circ. 1312). The capacity and application rates of the foam and extinguishing system shall comply with chapter 11 of the International Bulk Chemical Code, except that lower application rates may be accepted based on performance tests. For tankers fitted with inert gas*

systems, a quantity of foam concentrate sufficient for 20 minutes of foam generation may be accepted (refer to IMO MSC/Circ. 553).

Example calculation of the capacity of foam systems for oil tankers

Ship particulars:

- Beam = 14.5 m
- Length of cargo area = 56 m
- Length of largest cargo tank = 9 m
- Cargo deck area = 14.5 m × 56 m = 812 m^2
- Horizontal sectional area of single largest tank = 14.5 m × 9 m = 130.5 m^2

(Note: For the purpose of this illustration, a single tank encompasses the entire beam of the vessel).

- Proposed monitor spacing = 9 m

Area protected by largest monitor = 9 m × 14.5 m = 130.5 m^2

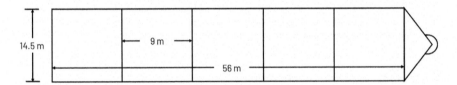

Figure 6.1 Calculation of foam system capacity – typical oil tanker.

Calculations based on the above criteria

(1) Determination of foam supply rate:
    (a) The largest of:
        The foam supply rate based upon the entire cargo deck area:
        0.6 l/min/m^2 × 812 m^2 = 487 l/min
    (b) The foam supply rate based upon the horizontal sectional area of the single largest tank:
        6.0 l/min/m^2 × 130.5 m^2 = 783 l/min
    (c) The foam supply rate based upon the area protected by the largest monitor:
        3.0 l/min/m^2 × 130.5 m^2 = 392 l/min[1]
        The foam supply rate is therefore 1,250 l/min, which is the largest of the three above calculated rates.

(2) Determination of the required quantity of foam concentrate:
1,250 l/min is the foam supply rate from (c). This flow rate for 20 minutes will require 20 min × 1,250 l/min = 25,000 l of foam-water solution.
If a 6% foam concentrate is used, then 6% of the 25,000 l must be foam concentrate, or 0.06 × 25,000 = 1,500 l.

(3) Determination of the minimum monitor capacity:
Each monitor must supply at least:
(a) 50% of the required system foam rate; or
(b) 3 l/min/m^2 for the area it protects; or
(c) 1,250 l/min, whichever is greater
50% of the foam supply rate = 1,250 l/min × 0.5 = 625 l/min
3.0 l/min/m^2 times the area the monitor protects = 130.5 m^2 × 3.0 l/min/m^2
= 390.6 l/min

Based on (c) above, the minimum monitor capacity is therefore 1,250 l/min.

Additional requirements for chemical carrier deck foam systems

All chemical carriers, except those solely dedicated to the carriage of non-flammable cargoes, must be fitted with a deck foam system complying with the specific Class rules pertaining to chemical-carrying vessels. Accordingly, a deck foam system should be provided which is capable of discharging foam at the required rates to the entire cargo tank deck area, as well as into any cargo that is open due to a rupture in the deck. This system will normally consist of fixed monitors, as well as required hand line applicators.

Controls

The deck foam system should be capable of simple and rapid operation. This would require that the controls and arrangements for the system are designed so that it can be brought into operation quickly and would preclude complicated interconnections with non-firefighting systems. The main control station for the system should be located outside the cargo area, adjacent to the accommodation spaces, and readily accessible and operable in the event of fire in the area being protected by the foam system. The deck foam system concentrate tank, proportioner unit, etc. are also normally installed in the above location but are not specifically mandated to be located at the control station. The pump(s) supplying water are normally located down in the engine room but the controls for the pumps, as well as all portions of the foam system not located within the space containing the foam system control station, should be operable from the foam system control station.

Application rate

The minimum flow rate for the deck foam system is determined by the appropriate Class rules. These typically identify three different sets of criteria, with the minimum capacity of the system required to be not less than the largest of the coverage rates identified. Typically, these different sets of criteria are:

- The requirement for a system flow rate of not less than 2.0 l/min/m^2 (0.049 gpm/ft^2) based on the total cargo tank deck area. The cargo tank deck area is further defined as the maximum breadth of the vessel multiplied by the total longitudinal extent of the cargo tank spaces.
- The requirement for a system flow rate of not less than 20.0 l/min/m^2 (0.49 gpm/ft^2) based on the horizontal deck area of the single largest cargo tank.
- The requirement for a system flow rate of not less than 10.0 l/min/m^2 (0.245 gpm/ft^2) of the area protected by the largest monitor, but not less than 1,250 l/min.

As indicated earlier, the minimum flow rate of the system must not be less than the largest rate identified in the three sets of criteria specified in the Class rules.

Foam concentrate quantity

The amount of foam concentrate carried onboard should be sufficient to supply the system for a period of at least 30 minutes when operating at the system's maximum flow rate, as determined earlier. The amount of foam concentrate required to be carried should be calculated based upon the maximum rated capacity of the system for the time period specified earlier.

System arrangements

The cargo deck foam system must be a fixed system and is to consist of monitors as well as foam applicators.

Foam monitors

The cargo deck foam monitors are a critical component of the deck foam system since they provide the crew with the ability to engage the fire by placing large quantities of foam where needed from a relatively safe location. The Class rules for the specific vessel in question will provide the requirements concerning the number, capacity, and locations of the monitors. Details of the monitor(s), as well as the proposed locations, should be regularly checked. As these requirements are generally shared by all Class administrations, these requirements are summarised here. Each monitor must have the capacity to deliver no less than 50% of the foam required by Class. In addition, each monitor should have a capacity of at least 10.0 l/min/m² (0.245 gpm/ft²) of the area protected by that monitor, but in no case less than 1,250 l/min (330 gpm). The capacity of the monitors should be verified by reviewing the monitor capacity curves produced by the manufacturer at the pressure indicated in the foam piping system pressure drop calculations. Accordingly, pressure drop calculations for the foam piping distribution system should be requested and reviewed. The number and position of the monitors must be adequate to provide coverage to the entire cargo tank deck area, as well as into any cargo tank that is open due to a rupture in the deck. The coverage area of any specific monitor is based on the following criteria:

- Only the area forward of the monitor is considered protected by that specific monitor.
- The distance to the farthest point of the coverage of a monitor is not to be more than 75% of the monitor throw in still air.
- Monitors are to be provided on both the port and starboard sides of the vessel at the front of the accommodations space.

Compliance regarding the locations and capacities of the monitors with the above requirements should be verified by reviewing a deck foam system arrangement drawing in conjunction with the monitor capacity and throw data provided by the monitor manufacturer.

Foam applicators

The foam applicators are also a critical component of the deck foam system as they provide the crew with the versatility and flexibility needed to combat a cargo tank fire. The Class rules for the specific vessel in question will provide the correct requirements concerning the number, capacity, and locations of the applicators. Details of the applicators as well as the proposed locations should be regularly verified. As these requirements are generally shared by all Class administrations, these requirements are summarised here. Each applicator must have the capacity to deliver no less than 400 l/min (105.7 gpm) and a throw of at least 15 m (49.2 ft) in still air. Once again, the capacity and throw of the applicators should be verified by reviewing the capacity and throw curves produced by the manufacturer at the pressure indicated in the foam piping system pressure drop calculations. Each monitor station should be provided with a hose connection for an applicator, including the monitor stations provided at the port and starboard sides of the vessel at the front of the accommodation space. The number and locations of the foam main outlets for hose connections to serve the foam applicators are to be sufficient so that at least two applicators can be directed to any part of the cargo tank deck area. At least four applicators with the appropriate hoses, connection devices, etc., should be carried onboard.

Isolation valves

The foam main should be fitted with isolation valves immediately forward of any monitor position so as to provide the means to isolate damaged sections of the main, as required by Class. These isolation valves should be of a type, design, and material such that they will not be rendered ineffective by heat.

Simultaneous operation

The operation of the deck foam system at its required output should permit the simultaneous use of the minimum required number of jets of water at the required pressure from the fire main. This is especially important to note when the fire pumps are being utilised to also supply the foam system. If the fire pumps are used for this service, the combined capacity of the fire pumps must be capable of supplying the total required fire main capacity at the required pressures while simultaneously supplying the maximum capacity required by the foam system.

Foam concentrate, quality, and testing

It is critical to recognise that the foam to be used must be suitable for the specific cargoes to be carried. As many chemical cargoes are polar solvents, the suitability of the foam for such cargoes should be verified. Also, it is important to recognise that the effectiveness of any specific foam concentrate may vary when used on fires from different chemicals. However, provisions of a system to supply multiple types of foam concentrates would be difficult and may introduce unwanted complications. Accordingly, the Class rules will indicate whether only one type of foam concentrate should be supplied, provided it is effective for the maximum possible number of cargoes intended to be carried. However, for other cargoes for which foam is not effective or is incompatible, additional arrangements to the satisfaction of the administration should be provided. Confirmation of the suitability of the foam concentrate with all chemicals to be carried is important.

Suitability of foam concentrate

As the quality of the foam and its suitability for use in the intended application are critical, the IMO has developed performance testing criteria for foam concentrates. This testing criterion is provided in IMO MSC.1/Circ.1312, '*Revised guidelines for the performance and testing criteria, and surveys of foam concentrates for fixed fire extinguishing systems*'. The testing procedure covers a wide spectrum of requirements, from freezing and drain time testing to expansion ratio and fire testing. Certain portions of the testing, such as the expansion ratio, drain time, and fire tests, are dependent upon the types of mixing and expansion equipment used (proportioners, monitors, etc.). Accordingly, it should be verified that the foam concentrate has been tested in accordance with IMO MSC.1/Circ.1312 and further that the equipment being provided onboard is the same equipment utilised during the testing or is the equipment that will produce foam which is equivalent to that tested.

Example calculation of the capacity of foam systems for chemical tankers

Ship particulars:

- Beam = 14.5 m
- Length of cargo area = 56 m
- Length of largest cargo tank = 9 m
- Cargo deck area = 14.5 m × 56 m = 812 m²
- Horizontal sectional area of single largest tank = 14.5 m × 9 m = 130.5 m²

Note: For the purpose of this illustration, a single tank encompasses the entire beam of the vessel).

- Proposed monitor spacing = 9 m
- Area protected by largest monitor = 9 m × 14.5 m = 130.5 m²

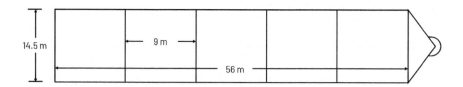

Figure 6.2 Calculation of foam system capacity for a typical chemical tanker.

Calculations based on the above criteria

(1) Determination of foam supply rate:
   The largest of:
   (a) The foam supply rate based upon the entire cargo deck area.
      2/min/m² × 812 m² = 1,624 l/min
   (b) The foam supply rate based upon the horizontal sectional area of the single largest tank:
      20 l/min/m² × 130.5 m² = 2,610 l/min

(c) The foam supply rate based upon the area protected by the largest monitor:

0 l/min/m^2 × 130.5 m^2 = 1,305 l/min^2

The foam supply rate is therefore 2,610 l/min, which is the largest of the three above calculated rates.

(2) Determination of the required quantity of foam concentrate:

2,610 l/min is the foam supply rate. This flow rate for 30 minutes will require:

30 min × 2,610 l/min = 78,300 l of foam-water solution.

If a 5% foam concentrate is used, then 5% of the 78,300 l must be foam concentrate, or

0.05 × 78,300 = 3,915 l.

(3) Determination of the minimum monitor capacity:

Each monitor must supply at least:

(a) 50% of the required system foam rate; or

(b) 10 l/min/m^2 for the area it protects; or

(c) 1,250 l/min, whichever is greater.

50% of the foam supply rate = 2,610 l/min × 0.5 = 1,305 l/min

10 l/min/m^2 times the area the monitor protects = 130.5 m^2 × 10 l/min/m^2

= 1,305 l/min

Based on (c) above, the minimum monitor capacity is therefore 1,305 l/min.

Additional requirements for fixed high expansion foam systems in machinery spaces

It is broadly recognised that fixed high expansion foam system is an acceptable system to meet the requirement for a fixed fire extinguishing system in Category 'A' machinery spaces. As such, it is generally a requirement that the high expansion foam system is a fixed system and capable of rapidly discharging the foam through fixed nozzles. The arrangement and distribution of the nozzles should be regularly reviewed to confirm adequate coverage of the protected space.

Application rate

The discharge rate for the high expansion foam system must be sufficient to fill the greatest space being protected at a rate of not less than 1 m (3.3 ft) per minute. Calculations, along with details of the system capacity, should be submitted to verify compliance with this requirement.

Foam concentrate quantity

The quantity of foam concentrate should be sufficient to produce a volume of foam equal to at least five times the volume of the largest space to be protected. The quantity of foam concentrate to be provided, in association with the particular foam's expansion ratio, should be reviewed and determined sufficient during plan review.

Foam concentrate

Adequate details of the foam should be submitted to confirm its suitability for the intended service. In addition, the details should identify the expansion ratio of the foam which is not to exceed 1,000:1.

System design

The supply ducts for delivering foam, the air intakes to the foam generator, and the number of foam-producing units should be such as to provide effective foam productions and distribution. Accordingly, the capacities of the foam generators, designs of the ducts, etc. should be reviewed to verify their suitability for the required system capacity.

System arrangements

The arrangement of the foam generator delivery ducting should be such that a fire in the protected space will not affect the foam-generating equipment. In addition, it should be noted that the location of the foam-generating equipment, foam concentrate storage, source of power (as well as the power supply arrangements), ventilation, source of water, etc., are to be such that a fire in the protected space will not affect the foam-generating system and the same should be verified during plan review. In association with the above, the location of the controls, as well as the foam-generating equipment, foam concentrate storage and source of power should be such that access to the spaces containing this equipment will not be cut off due to a fire in the protected space.

Foam concentrate, quality, and testing

The high expansion foam concentrate must be approved in accordance with IMO MSC/Circ.670, '*Guidelines for the performance and testing criteria and surveys of high expansion foam concentrates for fixed fire extinguishing systems*'. Like the testing requirements of IMO MSC.1/Circ.1312, '*Revised guidelines for the performance and testing criteria, and surveys of foam concentrates for fixed fire extinguishing systems*' discussed earlier, the testing procedures of IMO MSC/Circ.670 cover a wide spectrum of requirements, from freezing and drain time testing to expansion ratio and fire testing. Also, as certain portions of the testing, such as the expansion ratio, drain time, and fire tests, are dependent upon the types of mixing and expansion equipment used (proportioners, aerators, etc.), the equipment being provided onboard must be the same equipment utilised during the testing or equipment that will produce foam, which is equivalent to that tested.

Additional requirements for supplementary fixed low expansion foam systems in machinery spaces

The use of a fixed low expansion foam system is not considered acceptable as the required fixed fire extinguishing system in a Category 'A' machinery space. However, it is recognised that such systems are installed as a supplementary or additional system to fight fires in the bilge areas and may be relied on by the crew during a hazard. Accordingly, in addition to the general Class requirements noted earlier, specific Class guidance may provide additional criteria for the use of these types of systems. Where applicable, these requirements are outlined below.

System arrangement

The system should be arranged to discharge the foam through a permanent piping distribution system, fixed nozzles, and adequate control valves, which will allow for the effective distribution of the foam over the protected area(s).

System capacity

The system should be capable of generating sufficient foam in a five-minute period to cover the largest space over which oil fuel is liable to spread to a depth of 150 mm (6.0 in).

Expansion ratio

Adequate details of the foam should be submitted to confirm its suitability for the intended service. In addition, the details should identify the expansion ratio of the foam that is not to exceed 12:1.

Controls

The control arrangements for the system must be readily accessible, in a location that will not be cut off by a fire in the area being protected, and should be easy to operate. These arrangements should be verified during plan review, or if such information is not provided in the submission, a note is to be made requesting the Class Surveyor to verify the arrangements.

Foam concentrate, quality, and testing

The low expansion foam concentrate should be approved in accordance with IMO MSC.1/Circ.1312 '*Revised guidelines for the performance and testing criteria, and surveys of foam concentrates for fixed fire extinguishing systems*'. As discussed earlier, it is important to confirm that the equipment being provided onboard is the same equipment utilised during the testing or is the equipment that will produce foam that is equivalent to that tested.

Additional requirements for helicopter landing facilities

Application

Each helicopter deck onboard a vessel designated for helicopter operations must be provided with a fixed foam fire extinguishing system complying with the Class rules for that use. For clarification, the ABS Class rules indicate that a helicopter deck is considered a helicopter landing area on a vessel, where this includes all structures, firefighting appliances, and other equipment necessary for the safe operation of helicopters.

System arrangement

A fixed foam fire extinguishing system, consisting of monitors or hose streams or both, must be installed to protect the helicopter landing area.

System capacity

The system must be capable of delivering foam solution at a discharge rate in accordance with the following table for at least five minutes, and calculations verifying compliance with the same should be submitted and reviewed.

Table 6.1 System Capability for Helicopter Landing Facilities

Category	Helicopter overall length, L_H	l/min	gpm
H1	$L_H < 15$ m (49 ft)	250	66
H2	15 m (49 ft) $\leq L_H < 24$ m (79 ft)	500	132
H3	24 m (79 ft) $\leq L_H < 35$ m (11 5ft)	800	211

Foam concentrate

The foam agent is required to meet the performance standards for Level 'B' foam in the *International Civil Aviation Organisation's Airport Services Manual* (Part 1, Chapter 8, Paragraph 8.1.5, Table 8.1) and be suitable for use with seawater. Adequate details regarding the foam concentrate should be submitted to verify compliance with this requirement.

Simultaneous operation

The operation of the foam system must not interfere with the simultaneous operation of the fire main. Accordingly, where the foam system is supplied off the fire main system, the capacity of the fire pumps should be increased accordingly.

Note: Resolution MSC.403(96), which was formally adopted on 19 May 2016, and entered into force on 1 January 2020, includes a new chapter, Chapter 17 of the FSS Code, *Helicopter facility foam firefighting appliances*.

As we have seen throughout this discussion on the use of foam as a firefighting media, it is highly effective especially when deployed against chemical and oil-fuelled fires. Foam provides a substantially safer alternative to water, which is both conductive of electricity and can cause stability issues when large volumes of free surface water are introduced to higher levels of the vessel's superstructure. This problem is magnified further on vessels with large open decks, such as ROROs and ferries. Some vessels require specialist firefighting systems due to the nature of their cargoes. Gas carriers and oil tankers are two examples which immediately spring to mind, and which will form the focus of the next chapter of this book.

NOTES

1 Must not be less than 1,250 l/min.
2 Must not be less than 1,250 l/min.

Chapter 7

Fire extinguishing systems on gas carriers

Gas carriers present several unique fire hazards. Therefore, the firefighting systems used must be carefully reviewed to confirm they are adequate for the dangers involved. In this chapter, we will examine these hazards and discuss the systems in place to respond to them should the worst happen.

The unique hazards associated with gas carriers include:

- Vapour release of cargo, leading to creation of gas clouds.
- Boiling Liquid Expanding Vapour Explosions (BLEVE).
- Liquid pool fires, where discharge of water would only increase the evaporation rate and intensify the fire.
- Jet fires.

When a gas is released to the atmosphere, it will burn if it is within its flammable range and if exposed to a source of ignition. Depending upon the conditions under which combustion takes place, some degree of overpressure will also occur due to the rapid expansion of the heated gas. A liquid spill or vapour cloud burning over open water will develop little overpressure due to the unconfined nature of its surroundings. However, ignition of vapour within an enclosed space rapidly creates an overpressure sufficient to burst the boundaries. In cases of partial confinement, such as what might occur among shore plants and equipment, ignition may produce an overpressure sufficient to cause substantial damage, thus escalating the hazard and its consequences. If a leakage of liquid or vapour occurs from a pipeline under pressure, it will burn as a jet and continue to burn as long as fuel is supplied.

A particularly destructive form of vapour burn, BLEVE is associated with the storage of liquefied gas in pressurised containers. The BLEVE is a phenomenon associated with the sudden and catastrophic failure of a pressurised containment vessel when subjected to surrounding fire. This is one of the most devastating of liquefied gas accident scenarios. In all BLEVE incidents, the pressure vessel is subjected to flame impingement. BLEVE occurs when the fire increases internal tank pressure, and particularly at that part of the vessel not cooled by the internal liquid, the structure can be weakened to

DOI: 10.1201/9781003385523-9

the point of failure. As a result, the tank may suddenly split with fragments of the vessel's shell acting as projectiles. These projectiles are often expended with extreme force. Upon rupture, the sudden decompression produces a blast and the pressure immediately drops. At this time, the liquid temperature is well above its atmospheric boiling point, which then spontaneously boils off, creates large quantities of vapour that travels upward along with the liquid droplets. When the gas/air mixture is within its flammable limits, it will ignite from the tearing metal of the surrounding fire, creating a fireball that can reach gigantic proportions. A sudden release of gas provides further fuel for the rising fireball. The rapidly expanding vapour produces a further blast and intense heat radiation.

Small leaks from pump glands, pipe flanges, or from vent risers will initially produce vapour. This vapour will not ignite spontaneously. However, if the escape is large, then there may be a risk of the vapour cloud spreading to a source of ignition. If ignition does occur, it will almost certainly flash back to the leak. Leaks from pipelines are likely to be under pressure and, if ignited, will give rise to a jet flame. While arrangements for the emergency shut down of the pumps and remote closure of ESD valves are required, pressure may remain in a closed pipeline until the liquid trapped within has been expelled through the leak. In such a case, the best course of action is often to allow the fire to burn out. The alternative of extinguishing the fire has a high risk of producing a vapour cloud and having a flash back from re-ignition of the vapour. While the fire is being allowed to burn itself out, the surroundings should be protected with cooling water.

Significant pool fires are not likely on the vessel's decks because the amount of liquid which can be spilled in such a location is limited. The arrangement of the vessel's deck, with its camber and open scuppers, will allow liquid spillage to flow quickly and freely away over the vessel's side. Prompt initiation of ESD procedures further limits the availability of liquid cargo. However, any spillage of flammable gases in a liquefied state will result in a gas vapour cloud as the liquid evaporates. The gas generation rate will be large due to the low liquefaction storage temperature and the large amount of heat available from the surrounding structure and environment. Water should never be applied to a burning liquefied gas pool, which would provide a heat source for more rapid vapourisation of the liquid and increase the rate of burning. Any ignition of the ensuing vapour cloud would then result in a pool fire. The emissive power of a flame surface increases with pool diameter. LNG vapours burn in the initial stages with a comparatively clear flame; LPG, however, burns with a greater production of soot and, therefore, maximum surface emissive powers are lower than for LNG. Heat radiation levels from both LNG and LPG pool fires dictate that unprotected personnel must escape from the immediate vicinity as quickly as possible. Because of the damage which radiation can inflict on surrounding tanks and structure, such items are required to be protected by a water deluge system.

GENERAL PRINCIPLES OF CARGO DECK DRY CHEMICAL EXTINGUISHING SYSTEMS

Vessels carrying liquefied gases in bulk are required to be fitted with a dry chemical fire extinguisher conforming to IMO regulations. The system is used to protect the cargo deck area and all loading-station manifolds on the vessel. A deck system is typically made up of several independent skid-mounted units. The units are self-contained firefighting systems that use a dry chemical extinguishing agent propelled by a high-pressure inert gas such as nitrogen. Dry chemical extinguishing agents are chemicals in powder form. Dry chemical powders, such as sodium bicarbonate, potassium bicarbonate, and urea potassium bicarbonate, can be very effective in extinguishing small LNG or LPG fires. Dry chemical powders are effective in dealing with gas fires on deck or in extinguishing jet fires from a holed pipeline. They have been used successfully in extinguishing fires at vent risers.

Extinguishing effects of dry chemical

Dry chemical agents extinguish fire to the greatest extent by breaking the combustion chain. Minor amounts of cooling, smothering, and shielding of radiant heat are also present.

Chain breaking

As discussed in the first chapter of this book, chain reactions are necessary for continued combustion. In these chain reactions, fuel and oxygen molecules are broken down by heat, and they recombine into new molecules, giving off additional heat. This additional heat breaks down more molecules, which then recombine and give off still more heat. The fire thus builds, or at least sustains itself, through reactions that liberate enough heat to set off other reactions. Dry chemical (and other agents such as the halogen) attacks this chain of reactions. It does so by reducing the ability of the molecular fragments to recombine and form additional radicals. It also combines with the fragments of fuel and oxygen molecules so that the fuel cannot be oxidised. Although the process is not completely understood, chain breaking is the most effective extinguishing action of dry chemical.

Cooling

No dry chemical exhibits any great capacity for cooling. However, a small amount of cooling takes place simply because the dry chemical is at a lower temperature than the burning material. Heat is transferred from the hotter fuel to the cooler dry chemical when the latter is introduced into the fire.

Smothering

When dry chemical reacts with the heat and burning material, some carbon dioxide and water vapour are produced. These dilute the fuel vapours and the air surrounding the fire. The result is a limited smothering effect.

Shielding of radiant heat

Dry chemical produces an opaque cloud in the combustion area. This cloud reduces the amount of heat that is radiated back to the heart of the fire (i.e. the opaque cloud absorbs some of the radiation feedback that is required to sustain the fire). Less fuel vapour is produced, and the fire becomes less intense. The following information is provided as general guidance and instruction regarding the Class requirements for fixed dry chemical fire extinguishing systems. However, reference should always be made to the Class rules applicable to the specific vessel concerned for the complete set of requirements.

Dry chemical powder fire extinguishing systems

Vessels intended to carry flammable products are required to be fitted with a fixed dry chemical powder type extinguishing system for the purpose of fighting fires on the deck in the cargo area and bow or stern cargo handling areas, if applicable.

Coverage area and arrangement

The arrangement and capacity of the system should provide the ability to deliver powder from at least two hand hose lines or combination monitor/hand hose lines to any part of the above-deck exposed cargo area including above-deck product piping. In addition, at least, one hand hose line or monitor should be situated at the after end of the cargo area. The coverage considered from fixed monitors should be limited to the following:

Table 7.1 Fixed Monitor Requirements

Capacity of fixed qmonitor	10 kg/s	25 kg/s	45 kg/s
Maximum permitted coverage distance	10 m	30 m	40 m

Hand hose lines have a maximum effective distance of coverage that is equal to the length of hose. The location of fixed monitors and hand line stations, as well as the hose lengths, should be re-evaluated to confirm compliance with the above requirements.

Dry powder units

Vessels with a cargo capacity of 1,000 m^3 (35,315 ft^3) or more require a dry chemical fire extinguishing system. This system must consist of at least two independent and self-contained, dry chemical powder units, which include the associated controls, pressurising medium fixed piping, monitors, and/ or hand hose lines. For vessels with a cargo capacity of less than 1,000 m^3 (35,315 ft^3), only one unit is necessary. This system must be activated by an inert gas such as nitrogen, used exclusively for this purpose and stored in pressure vessels adjacent to the powder containers. Details of the powder units should be requested and compliance with the same confirmed.

Pressure vessels

All pressure vessels associated with the powder units are required to comply with the Class rules for this type of equipment, and should have appropriate certification verified, as necessary.

Controls

If monitors are installed, they must be capable of actuation and discharge both locally and remotely. The monitor does not have to be aimed remotely, provided it can deliver the necessary powder (from a single position) to all necessary areas of coverage.

Piping arrangements

A fire extinguishing unit with two or more monitors, hand hose lines, or a combination thereof should have independent pipes (with a manifold at the powder container) unless a suitable alternative means is provided to give proper performance. Where two or more pipes are attached to a unit, the arrangement should be such that any or all the monitors and hand hose lines should be capable of simultaneous or sequential operation at their rated capacities. Where fixed piping is provided between the powder container and a hand hose line or monitor, the length of piping should not exceed that length, which can maintain the powder in a fluidised state during sustained or intermittent use, and which can be purged of powder when the system is shut down. This can be difficult to verify since the ability to maintain the powder in solution is dependent on several issues. As a minimum, appropriate criteria from the manufacturer justifying the sizing and length of piping should be requested.

Monitors and hand lines requirements

The capacity of a monitor should not be less than 10 kg/s (22 lbs/s). Hand hose lines should be kink proof and fitted with a nozzle capable of on/ off operation and discharge at a rate not less than 3.5 kg/s (7.7 lbs/s). The maximum discharge rate should be such as to allow operation by one person. The length of a hand hose line should not exceed 33 metres (108.3 feet). Hand hose lines and nozzles should be of weather-resistant construction or stored in weather-resistant housing or covers and be readily accessible. These requirements will be provided in the appropriate Class rule document. In addition, the monitors should be certified in accordance with Class requirements.

System capacity

Enough dry chemical powder should be stored in each container to provide a minimum of 45 seconds of discharge time for all monitors and hand hose lines attached to each powder unit.

Bow and stern loading and discharge arrangements

Vessels fitted with bow or stern loading and discharge arrangements must be provided with an additional dry chemical powder unit complete with at least one monitor and one hand hose line. This additional unit should be located to protect the bow or stern loading and discharge arrangements. In addition, hand hose lines are to be provided to protect the area of the cargo line forward or aft of the cargo area.

General piping requirements

In addition to the above, system piping components are to comply with the general requirements and limitations set by Class. In addition, depending upon the pressure 'Class' of the system, the components may also be required to be certified in accordance with specific Class rules.

CARGO AREA WATER SPRAY SYSTEMS

Fixed water deluge systems are required to be provided to protect surfaces such as vessel's structures, cargo tanks, and piping, which can be exposed to liquefied gas fires. Such systems are designed to provide a layer of water over the exposed surfaces, and by this means, the radiant heat from the fire is absorbed by the water. Provided a water layer of some thickness can be maintained, the surface temperature usually will not exceed 100 °C (212 °F).

Areas to be protected

Vessels carrying flammable or toxic products should have a water spray system for cooling, fire prevention, and crew protection equipment that can cover the following:

- Exposed cargo tank domes, any exposed parts of cargo tanks, and any part of cargo tank covers that may be exposed to heat from fires in adjacent equipment containing cargo such as exposed booster pumps/heaters/re-gasification or re-liquefaction plants, addressed as gas process units, positioned on weather decks.
- Exposed on-deck storage containers of flammable or toxic products.
- Gas process units positioned on deck.
- Cargo liquid, vapour discharge, and loading connections, including the presentation flange and the area where their control valves are situated, which shall be at least equal to the area of the drip trays provided.
- All exposed emergency shut-down (ESD) valves in the cargo liquid and vapour pipes, including the master valve for supply to gas consumers.
- Exposed boundaries facing the cargo area, such as bulkheads of superstructures and deck houses normally manned.
- Cargo machinery spaces, storerooms containing high fire risk items, and cargo control rooms. Exposed horizontal boundaries of these areas do not require protection unless detachable cargo piping connections are arranged above or below.

Note: Boundaries of unmanned forecastle structures that do not contain high fire risk items or equipment do not require water spray protection.

- Exposed lifeboats, life rafts and muster stations facing the cargo area, regardless of distance to cargo area; and any semi-enclosed cargo machinery spaces and semi-enclosed cargo motor room.

Coverage rate

The system should be capable of covering all areas mentioned with a uniformly distributed water spray of at least 10 l/min/m^2 (0.24 gpm/ft^2) for horizontal projected surfaces and 4 l/min/m^2 (0.096 gpm/ft^2) for vertical surfaces. For structures having no clearly defined horizontal or vertical surfaces, the capacity of the water spray system should be the greater of the projected horizontal surface multiplied by 10 l/min/m^2 (0.24 gpm/ft^2) or the actual surface multiplied by 4 l/min/m^2 (0.096 gpm/ft^2). On vertical surfaces, the spacing of nozzles protecting lower areas may account for anticipated run down from higher areas.

Piping arrangement

For isolating damaged sections, stop valves are required to be fitted in the spray system at various intervals. Alternatively, the system is permitted to be divided into two or more sections, which may be operated independently provided that the necessary controls are located together aft of the cargo area. However, any section protecting special areas, for example, exposed cargo tank domes, any exposed parts of cargo tanks, and exposed on-deck storage vessels for flammable or toxic products should cover the whole of the athwartship tank grouping which includes that area. In addition, the vertical distances between water spray nozzles for protection of vertical surfaces are not to exceed 3.7 m (12 ft).

Pumping capacity

The capacity of the water spray pumps should be sufficient to deliver the required amount of water to all areas simultaneously or where the system is divided into sections, the arrangements, and capacity should be such as to supply water simultaneously to any one section and to the surfaces specified earlier as indicated by Class. Alternatively, the main fire pumps may be used for this service, provided that their total capacity is increased by the amount needed for the spray system. In either case, a connection, through a stop valve, should be made between the fire main and water spray main outside the cargo area. Water pumps normally used for other services may be arranged to supply the water spray main if specifically approved by the administration.

General piping requirements

In addition to the standardised requirements set by Class, all pipes, valves, nozzles, and other fittings in the water spray systems should be resistant to corrosion by seawater (for which purpose galvanised pipe, e.g., may be used) and to the effect of fire.

Controls

Remote starting of pumps supplying the water spray system and remote operation of any normally closed valves in the system should be arranged in suitable locations outside the cargo area, adjacent to the accommodation spaces and readily accessible and operable in the event of fire in the areas protected. Details verifying compliance with the above should be regularly reviewed.

Fire main system

Refer to the beginning of the chapter for more information about the requirements for fire main systems installed on gas carriers.

ENCLOSED SPACES CONTAINING CARGO HANDLING EQUIPMENT

Enclosed spaces meeting the criteria of cargo machinery spaces, and the cargo motor room within the cargo area of any ship, are to be provided with a fixed fire extinguishing system complying with the provisions of the FSS Code and accounting for the necessary concentrations and/or application rates required for extinguishing gas fires. Enclosed spaces meeting the criteria of cargo machinery spaces within the cargo area of ships that are dedicated to the carriage of a restricted number of cargoes are to be protected by an appropriate fire extinguishing system for the cargo carried. Turret compartments of any ship are to be protected by internal water spray, with an application rate of not less than 10 $l/m^2/min$ of the largest projected horizontal surface. If the pressure of the gas flow through the turret exceeds 4 MPa, the application rate is to be increased to 20 $l/m^2/min$. The system should be designed to protect all internal surfaces

Chapter 8

Portable and semi-portable fire extinguishers

Given a fire usually starts small, most fires that are discovered early and attacked quickly can usually be controlled and extinguished before they grow out of control. The ready availability of suitable portable and semi-portable fire extinguishers is therefore very important. Although limited in capacity, portable extinguishers are easy to transport and can be used to engage a fire quickly. Semi-portable extinguishing systems bring larger amounts of extinguishing media (sometimes called agent) to the fire but are more difficult to transport. When used properly, both can be very effective in controlling and extinguishing small, localised fires. With that said, not all extinguishers are the same. Fire extinguishers vary in terms of their size and by the extinguishing media that they use. Importantly, not every extinguisher is suitable for every type of fire. Depending on the type of media, some extinguishers can exacerbate a fire or worse, cause injury to the person using the extinguisher. It is therefore critical that the correct extinguisher is used on the right type of fire.

The usual types of extinguishing media found in portable extinguishers onboard seagoing vessels include:

- Water.
- Foam.
- Carbon dioxide (CO_2).
- Dry powder.
- Dry chemical.
- Wet chemical.

Portable extinguishers using Halon gas as the extinguishing media have also been very popular in the past. However, with the Montreal Protocol and the restrictions developed regarding the manufacture and use of Halon, such extinguishers have been phased out and so will not be covered in this book. Within the various types of extinguishing media, there are differences between the specific extinguishing media and the methods by which that

DOI: 10.1201/9781003385523-10

media is expelled. Extinguishers also vary significantly in size ranging from very small portables, which can be carried to the location of the fire quickly and easily, to large semi-portable units, which consist of a large tank on wheels, a reel, and long hose.

Note: This chapter examines portable and semi-portable fire extinguishers only. For more information about other types of portable firefighting appliances required by Class rules (e.g. portable foam applicators, fire axes, and fireman's outfits), refer to Chapter 10, *Fire control plans*.

PORTABLE AND SEMI-PORTABLE FIRE EXTINGUISHERS

The following sections discuss the types of portable extinguishers available and their differences within the individual categories of extinguishers. A general description of the various types of portable extinguishers (including individual comparisons, characteristics, and limitations) is provided in IMO Resolution A.951(23). For more information, refer to the FSS Code, Chapter 4, and the appendix to IMO Resolution A.951(23).

Water-type fire extinguishers

Extinguishers that use water or water solution as the extinguishing media are suitable only for class 'A' fires (refer to Chapter 2 for more information about fire classifications). There are typically three types of water extinguishers used in shipboard settings, which are soda acid, cartridge operated, and stored pressure water extinguishers. The manufacture of soda acid and cartridge-operated extinguishers has generally been discontinued; however, since large numbers of these types of extinguishers are still in use, they will be discussed here with the most common type, the stored pressure water extinguisher. Water extinguishers may be filled with either water or a water and antifreeze solution. Accordingly, water extinguishers could be subject to freezing, and adherence to the manufacturer's recommendations regarding the environmental conditions of the storage locations is important.

Soda acid extinguishers

The soda acid extinguisher normally comes with a 9.5 l (2.5 gal) size tank and weighs approximately 13.6 kg (30 lbs) when charged. It has a typical reach of between 10.7 and 12.2 m (30 to 40 ft), and expends itself in about 55 seconds. The shell of the extinguisher is filled with a solution of 0.7 kg (1.5 lbs) of sodium bicarbonate in 9.5 l (2.5 gal) of water. The screw-on cap contains a cage that holds a 0.23 kg (8.1 oz) bottle, half filled with sulphuric acid, in an upright position. A loose stopper in the top of the acid bottle prevents acid from splashing out before the extinguisher should be used. The extinguisher is carried to the fire by means of the top handle. At the fire, the extinguisher is inverted; the acid mixes with the sodium bicarbonate solution, which then forms carbon dioxide gas. The pressure of the CO_2 can then propel the water out through the nozzle. The stream should be directed to the seat of the fire and then moved back and forth to hit as much of the fire as possible. Typically, there is no discharge control mechanism; the nozzle should be directed at the fire until the entire content of the extinguisher is discharged.

Important! The extinguishing agent, sodium bicarbonate solution, when mixed with acid, is corrosive and it should be recognised that extreme care should be taken to avoid contact with the agent on the skin or face, as this acid can cause severe burns.

It is equally as important that soda acid extinguishers are properly maintained. This type of extinguisher is not normally kept under pressure. Therefore, when the extinguisher is inverted for use, a pressure of 896 kPa (130 psi) or more is suddenly generated. If the extinguisher is not properly maintained, and/or the container is corroded or otherwise damaged, then the sudden pressure could be sufficient to burst the container.

Cartridge-operated extinguishers

There are two types of cartridge-operated extinguishers: the rupture disc-type and the pin-type.

Rupture disc-type, cartridge-operated extinguisher

The cartridge-operated water extinguisher is similar in size to the soda acid extinguisher. The most common size is 9.5 l (2.5 gal) and has a range of about 10.7 to 12 m (30 to 40 ft). The container is filled with water or an antifreeze solution. The screw-on cap contains a small cylinder of CO_2, and when the CO_2 cylinder is punctured, the gas provides the pressure to propel the extinguishing agent. For use, the extinguisher is carried to the fire, and then inverted and bumped against the deck. This ruptures the CO_2 cylinder and expels the water. Once again, the stream should be directed at the seat of the fire. The nozzle should be moved back and forth to quench as much of the burning material as possible in the short time available. The discharge time is less than one minute, and the entire content of the extinguisher will be discharged, since like the soda-acid extinguisher, typically the flow cannot be shut off. As with the soda-acid extinguisher, the container is not subjected to pressure until it is put to use. Thus, any weakness in the container may not become apparent until the container fails.

Pin-type cartridge-operated extinguisher

A newer version of the cartridge-operated water extinguisher is the pin-type cartridge-operated extinguisher that does not need to be inverted for use. Instead, a pin is pulled out of the cartridge, with the extinguisher upright. A lever is squeezed to discharge the extinguishing agent (water or anti-freeze solution).

Stored-pressure water extinguishers

The stored-pressure water extinguisher typically comes in a 9.5 l (2.5 gal) size and weighs about 13.6 kg (30 lbs). It has a horizontal range of about 10.7 to 12.2 m (35 to 40 ft). In continuous operation, it expends its water in about 55 seconds. However, stored pressure water extinguishers are typically fitted with some means to control the discharge, and therefore, may be used intermittently, to extend their operational time. The container is filled with water or an antifreeze solution, to within about 15 cm (6 in) of the top. Most extinguishers have a fill mark stamped on the container. The screw-on cap holds a lever-operated discharge valve, a pressure gauge, and an automobile tyre-type valve. The extinguisher is pressurised through the air valve, with either air or an inert gas such as nitrogen. The normal charging pressure is about 690 kPa (100 psi). The gauge allows the pressure within the extinguisher to be checked at any time. Most gauges are colour coded to indicate normal and abnormal pressures. The extinguisher is carried to the fire, and the ring pin or other safety device is removed. The operator aims the nozzle with one hand and squeezes the discharge lever with the other hand. The stream should be directed at the seat of the fire and moved back and forth to provide complete coverage of the burning material. Short bursts can be used to conserve the limited supply of water.

Foam-type fire extinguishers

Foam extinguishers typically have greater range of extinguishing capability compared to water extinguishers. The most common size is a 9.5 l (2.5 gal) unit and may be used on both class 'A' and class 'B' fires. On class 'A' fires, the foam serves to cool the fire and fuel source, while on class 'B' fires, the foam serves to act as a barrier which excludes oxygen from the fuel surface below the ignition surface. Foam extinguishers typically have a range of between 6.1 and 7.7 m (20 to 25 ft) and a discharge duration of slightly less than one minute. There are two different types of foam extinguishers available:

- Chemical foam.
- Mechanical foam.

Note: Chemical foam extinguishers have been phased out, but may still be found on some shipboard settings.

Like water extinguishers, foam extinguishers may also be subject to freezing, and adherence to the manufacturer's recommendations regarding the environmental conditions of the storage locations is critical.

Chemical foam portable fire extinguishers

The chemical foam extinguisher is charged by filling it with two solutions that are kept separate (within the extinguisher) until it should be used. These solutions are commonly referred to as the 'A' and 'B' solutions. Their designations have nothing to do with fire classifications. The foam extinguisher is carried to the fire right side up and then inverted. This mixes the two solutions, producing a liquid foam and CO_2 gas. The CO_2 acts as the propellant and as the gas for forming the foam bubbles. The liquid foam expands to about eight times its original volume, meaning that a standard 9.5 l (2.5 gal) extinguisher will produce approximately 68 to 76 l (18 to 20 gal) of foam. The foam should be applied gently on burning liquids (class 'B' fires). This can be done by directing the stream in front of the fire. The stream may also be directed against the back wall of a tank or a structural member to allow the foam to run down and flow freely over the fire. Chemical foam is stiff and flows slowly. For this reason, the stream must usually be directed to the fire from several angles, for complete coverage of the burning materials. For fires involving ordinary combustible materials (i.e. class 'A' fires), the foam may be applied in the same way (as a blanket), or the force of the stream may be used to project the foam into the seat of the fire. Once activated, these extinguishers will expel their entire foam content which should be directed onto the fire. As with other pressurised extinguishers, the containers are subject to rupture when their contents are mixed and are a possible cause of injury to the operator.

Mechanical foam extinguishers

Mechanical foam portable extinguishers contain a pre-mixed solution of film-forming foam surfactant and water. The solution is stored under a pressurised charge of carbon dioxide or other gas which serves as the expelling force. The discharge hose is fitted with an air aspirating nozzle, which expands the solution into foam as it is discharged through the nozzle. Mechanical foam extinguishers closely resemble the pressurised water extinguishers except for the air aspirating nozzle. Like the chemical foam extinguisher, the foam should be applied gently on burning liquids. This can be done by directing the stream in front of the fire. The stream also may be directed against the back wall of a tank or a structural member to allow the foam to run down and flow over the fire.

Carbon dioxide (CO_2) fire extinguishers

Carbon dioxide extinguishers are used primarily on class 'B' and class 'C' fires. The most common size of portable extinguisher for shipboard use is the 5 kg (11 lbs) CO_2 extinguisher. This weight does not include that of the relatively heavy metal shell. The CO_2 is stored mostly in the liquid state at a pressure of 5,860 kPa (850 psi) at 21 °C (70 °F). The range of the CO_2 varies between 1.8 and 2.4 m (3 to 8 ft), and the duration lasts between 8 and 30 seconds. The extinguisher is carried to the fire in an upright position. The short range of the CO_2 extinguisher means the operator must get fairly close to the fire. The extinguisher is placed on the deck, and the locking pin is removed. The discharge is normally controlled either by opening a valve or by squeezing the two handles together. The operator must grasp the hose handle and not the discharge horn. The CO_2 expands and cools very quickly as it leaves the extinguisher. The horn gets cold enough to frost over and can cause severe non-freezing cold injuries. When a CO_2 extinguisher is used in a confined space, the operator should guard against suffocation by wearing breathing apparatus. For use on class 'B' fires, the horn should be aimed first at the base of the fire nearest the operator. The discharge should be moved slowly back and forth across the fire. At the same time, the operator should move forward slowly. The result should be a 'sweeping' effect of the flames off the burning surface, with some carbon dioxide 'snow' left on the surface. Whenever possible, a fire on the weather deck should be attacked from the windward side. This will allow the wind to blow the heat away from the operator and to carry the CO_2 to the fire. It should be noted that generally, CO_2 extinguishers do not perform well in windy conditions. The blanket of CO_2 gas does not remain on the fire long enough to permit the fuel to cool down. For use on class 'C' fires, the discharge should be aimed at the source of a fire that involves electrical equipment. The equipment should be de-energised as soon as possible to eliminate the chance of shock and the source of ignition. CO_2 extinguishers do not need to be protected against freezing. However, they should be stowed at temperatures below 54 °C (130 °F) to keep their internal pressure at a safe level. At about 57 °C (135 °F), the safety valves built into CO_2 extinguishers are activated at approximately 18,620 kPa (2,700 psi) to release excess pressure.

   Note: SOLAS Regulation II-2/10.3.2.3 indicates that CO_2 extinguishers shall not be placed in accommodation areas, as the inadvertent or undetected release of the CO_2 can lead to the suffocation of unsuspecting personnel.

Dry chemical-type fire extinguishers

Dry chemical extinguishers are available in several sizes and are supplied with any of five different extinguishing medias. Each dry chemical agent has different extinguishing capabilities. If sodium bicarbonate is arbitrarily given an extinguishing capability of 1, then the relative capabilities of the other dry chemical agents are as follows:

- Mono-ammonium phosphate (ABC) 1.5.
- Potassium chloride (BC) 1.8.
- Potassium bicarbonate (BC) 2.0.
- Urea potassium bicarbonate (BC) 2.5.

Dry chemical extinguishers typically extinguish fires by inhibiting the chemical reaction of the fire process (refer to the section on the fire tetrahedron in Chapter 1). All five dry chemicals mentioned earlier are suitable for extinguishing class 'B' and class 'C' fires. However, one dry chemical extinguishing agent, mono-ammonium phosphate (ABC, multi-purpose), is also approved for use on class 'A' fires. This agent extinguishes the fire by chain breaking (as do the other dry chemical agents). As a by-product of this chemical reaction, a residue is created that coats and clings to the surface of the burning materials. This coating deprives solid fuels of air and has therefore been found to be acceptable for use on class 'A' fires. Since dry chemical extinguishers extinguish fires mainly through the breaking of the chemical chain reaction, there is little or no cooling of the flame or fuel surface. Thus, a reflash is possible if the fuel continues to remain in contact with hot surfaces. Additional dry chemical extinguishers or other appropriate extinguishers should be available as backup until all sources of ignition are eliminated. Dry chemical extinguishing agents may be used together with water and some dry chemical extinguishers are filled with an extinguishing agent that is compatible with foam. There are essentially two different types of dry chemical extinguishers: the cartridge-operated extinguisher and the stored-pressure extinguisher. The following discussion provides a description of each.

Cartridge-operated dry chemical extinguisher

Portable, cartridge-operated, dry chemical extinguishers range in size from 2.25 to 13.6 kg (5 to 30 lbs), and semi-portable models contain up to 22.7 kg (50 lbs) of agent. The extinguisher may be filled with any of the five agents, and its rating will be based on the specific agent employed. A small cylinder of inert gas is used as the propellant. Cartridge-operated, dry chemical extinguishers have a range of between 3 and 9.1 m (10 to 30 ft). The 2.25 kg (5 lbs) extinguisher will have a discharge duration of approximately 8 to 10 seconds, whilst larger extinguishers can provide upwards of 30 seconds of discharge time. The extinguisher is carried and used upright. The ring pin is removed, and the puncturing lever is depressed. This releases the propellant gas, which forces the extinguisher agent up through the nozzle. The flow of dry chemical is controlled with the squeeze-grip 'On-Off' nozzle at the end of the hose. The discharge is directed at the seat of the fire, starting at the near edge. The stream should be moved from side to side with rapid motions, to sweep the fire off the fuel. On the weather deck, the fire should be approached from the windward side wherever possible. The initial discharge should not be directed onto the burning material from close range, that is, between 0.91 and 2.4 m (3 to 8 ft), as the velocity of the stream may scatter burning material. However, the agent may be applied in short bursts by opening and closing the nozzle with the squeeze grips. If the propellant gas cylinder is punctured but the extinguisher is not put into use or is only partially discharged, the remaining gas may leak away in a few hours. Thus, the extinguisher must be recharged after each use or activation.

Stored pressure dry chemical extinguishers

Stored pressure dry chemical extinguishers are available in the same sizes as cartridge operated types. They have the similar ranges and duration of discharge and are used in the same way. The only differences are that the propellant gas is mixed in with the dry chemical in the stored pressure type and the extinguisher is controlled with a squeeze-grip trigger on the top of the container. A pressure gauge indicates the condition of the charge. Many stored pressure extinguishers have pressure gauges that indicate whether the internal pressure is within the operating range. The gauge is located on the bottom of some extinguishers.

Dry powder-type fire extinguishers

Frequently the terms 'dry powder' extinguisher and 'dry chemical' extinguisher are used, incorrectly, as interchangeable. The two terms represent very different types of medium, and it is important to understand that there is a difference between the two types of extinguishers. Dry chemical extinguishers and their mediums, as discussed earlier, are typically suitable for use on class 'B' and class 'C' fires, or in the case of mono-ammonium phosphate, suitable for use on class 'A', class 'B' and class 'C' fires. However, the extinguishing media in a 'dry powder' extinguisher is a special type of dry chemical agent that is specifically suitable for use on combustible metal (class 'D') fires. Accordingly, the term 'dry powder' extinguisher is intended to specifically refer to that extinguisher which has an extinguishing agent suitable for use on combustible metal (class 'D') fires. The extinguishing agent typically found in 'dry powder' extinguishers is sodium chloride, which forms a crust on the burning metal.

Wet chemical-type fire extinguishers

Wet chemical is a new agent that extinguishes the fire by removing the heat element of the fire tetrahedron and prevents re-ignition by forming a barrier between the oxygen and fuel elements. Wet chemical extinguishers predominantly contain a solution of potassium acetate, sometimes with some potassium citrate or potassium bicarbonate and are mainly used on class 'F' or 'K' fires (i.e. fires involving cooking grease, fats, and oils) while some may also be used on class 'A' fires. The most common size is a 6 l (1.5 gal) or 9.5 l (2.5 gal) unit, typically having a discharge range of 2.4 to 3.7 m (8 to 12 ft) and a duration of between 50 and 55 seconds (for the 6 l) or 75 and 80 seconds (for the 9.5 l). Class 'F' or 'K' fires most often occur in the vessel's galley where cooking media (fats, greases, and oils) are stored and used. Wet chemical extinguishers work on the principle of *saponification*. Saponification takes place when alkaline mixtures such as potassium acetate, potassium citrate, or potassium carbonate are applied to burning cooking oil or fat. The alkaline mixture combines with the fatty acid to create a soapy foam layer on the surface. This holds in the vapours and steam, extinguishing the fire. Recent changes to cooking operations have presented major challenges to dry chemical fire extinguishers and systems. Changes in frying oils from animal fats to vegetable oils have reduced the ability of dry chemicals to extinguish galley fires. Since vegetable oils have lower fatty acid content, many vegetable oils prevent the 'foam blanket' from developing completely. This inhibits the extinguishing agent from working properly by allowing vapours and steam to release. In addition, newer efficient fryers retain heat much longer than in the past. Moreover, vegetable oils have a much higher auto-ignition temperature than animal fats. Therefore, dry chemical agents have trouble preventing a re-flash from occurring and cannot pass the current test standards for fryers due to the presence of retained heat. Heat breaks down the weaker foam layer that is created, making it necessary to cool the oils in addition to the foam layer. To address these issues, wet chemical agents were introduced and are now required as per SOLAS regulation II-2/10.3, referring to IMO MSC.1/Circ.1275 'Unified Interpretation of SOLAS Chapter II-2 on the number and arrangement of portable fire extinguishers on board ships' for galleys with deep fat fryers.

Wet chemical-type extinguishers are alkaline in nature and are the only type of extinguishing agent approved for cooking area fire suppression. This is down mainly to their ability to maintain a sufficient foamy layer, which allows for complete cooling. In addition, wet chemical agents pose minimal damage threat to hot appliances, minimise splash hazards, offer improved visibility during firefighting, and provide for an easier clean up than is typically the case with dry chemical agents. The wet chemical extinguisher is carried to the fire and then held in an upright position. The pin is twisted to break the seal and pulled to release the trigger mechanism. The applicator is held and aimed at the base of the flames. The trigger is squeezed and gripped so that the upper and lower parts of the trigger are pressed together. This allows the agent of the extinguisher to discharge. During discharging, the operator should apply the wet chemical using the extended applicator in a slow circular movement, which gives a gentle, yet highly effective, application. It is important to apply the fine spray onto the burning fat until the surface of the burning cooking oil changes into a soap-like substance. This will prevent re-ignition. The gentle application also helps to prevent burning oil from splashing out of the container.

PORTABLE FIRE EXTINGUISHERS

Portable extinguishers are smaller extinguishers that can be easily carried to the site of the fire. IMO Resolution A.951(23) defines a portable extinguisher as one which is designed to be carried and operated by hand and which in a working order has a total weight of not more than 23 kg (50.7 lbs). For more information, refer to the appendix of IMO Resolution A.951(23). Although portable fire extinguishers provide only a limited quantity of fire extinguishing agent, their ability to be carried easily to a fire and used to quickly engage a fire before it spreads plays a vital role onboard any shipboard setting. Many potentially large-scale fires have been avoided simply due to the availability and effective use of small portable fire extinguishers.

SEMI-PORTABLE FIRE EXTINGUISHERS

Semi-portable extinguishers provide a larger amount of extinguishing agent to a fire rapidly, which allows the operator to make a sustained attack. However, semi-portable fire extinguishers are much larger, which results in a restriction of mobility. Semi-portable units may be wheeled units or may be semi-fixed and typically utilise a discharge hose, which can be run out to engage the fire. The hose must be of sufficient length to reach all portions of the protected area. Semi-portable systems typically utilise foam, carbon dioxide, dry chemical, or dry powder, depending upon the requirements of the hazard to be protected. They are often required to be provided at high fire risk areas within spaces which may already be fitted with a fixed fire extinguishing system. The purpose of the semi-portable units should provide the means to quickly engage a fire with larger volumes of extinguishing agent. If this attack controls or extinguishes the fire, then the large fixed system need not be activated. Semi-portable systems may also be used as primary extinguishing systems.

FIRE EXTINGUISHER DESIGNATIONS

Portable and semi-portable extinguishers are typically classified with one or more letters and with a numeral. The letter (or letters) indicates the classes of fires on which the extinguisher may be safely used. These letters correspond

Table 8.1 Portable and Semi-portable Extinguisher Classifications

Type	Size	Water l (gallon)	Foam l (gallon)	Carbon dioxide Kg (lbs)	Dry chemical kg (lbs)
A	II	9 (2$^1/_2$)	9 (2$^1/_2$)	—	5 (11)
B	II	—	9 (2$^1/_2$)	5 (11)	5 (11) 'ABC' only
B	III		45 (12)	15.8 (35)	9 (20)
B	V		152 (40)	45 (100)	22.5 (50)
C	I		—	1.8 (4)	0.9 (2)
C	II		—	5 (11)	5 (11)

exactly to the five classes of fires (refer to Chapter 2). Thus, for example, class 'A' extinguishers may be used only on class 'A' fires – that is, those involving common combustible materials. Class 'AB' extinguishers may be used on fires involving wood (class 'A') or diesel oil (class 'B'), or indeed both. The numeral indicates either the relative efficiency of the extinguisher or its size. The specific size and classification ratings for a specific manufacturer's extinguisher should be established through appropriate testing conducted by a recognised testing agency or laboratory. Most Class administrations use Roman numerals to indicate the sizes of portable extinguishers. The numeral 'I' indicates the smallest size, and 'V' the largest. Sizes 'I' and 'II' are hand portable extinguishers and sizes 'III' and 'V' are semi-portable extinguishers. The ratings of the different types of extinguishers are provided in Table 8.1.

CLASS REQUIREMENTS FOR PORTABLE AND/OR SEMI-PORTABLE EXTINGUISHERS

This section provides general guidance regarding the Class requirements for portable and/or semi-portable fire extinguishers. However, reference should always be made to the specific Class rules applicable to the vessel concerned for a complete set of current requirements.

Sizing of portable extinguishers

Establishment of a minimum size for portable extinguishers is important to provide adequate extinguishing media. Accordingly, the standardised Class rules establish a 9 l (2.5 gal) fluid extinguisher as a baseline for the extinguishing capability of a portable extinguisher and require any other extinguisher to have the fire extinguishing capability of not less than that provided by a 9 l (2.5 gal) fluid extinguisher. Therefore, CO_2 and dry chemical extinguisher must establish, through satisfactory testing, that they provide a fire extinguishing capacity equivalent to that of the 9 l (2.5 gal) fluid extinguisher for each specific fire hazard. The portability of the extinguisher is also a critical factor. IMO Resolution A.951(23) defines a portable extinguisher as one which is designed to be carried and operated by hand and, which in working order, has a total weight of not more than 23 kg (50.7 lbs). For further information, refer to IMO Resolution A.951(23). In accordance with IMO Resolution A.951(23), Class rules have adopted these same parameters to ensure parity with the FSS Code, Chapter 4/3.1.1.1, which states that the mass of portable fire extinguishers should not exceed 23 kg (50.7 lbs). Each powder or carbon dioxide extinguisher should have a capacity of at least 5 kg (11 lbs) and each foam extinguisher a capacity of at least 9 l (2.5 gal).

APPROVAL OF PORTABLE AND SEMI-PORTABLE EXTINGUISHERS

All portable extinguishers must be of an approved type and design (in the context of this requirement, the term 'portable extinguisher' should be considered to apply to both portable and semi-portable extinguishers). Therefore, the extinguisher should be designed and fabricated to an appropriate 'marine' portable extinguisher standard, which addresses the needs and concerns associated with the marine environment (i.e. corrosion). To verify compliance, the extinguisher should have documentation confirming its approval by an appropriate agency or laboratory for 'marine' service. This is in addition to the size and classification ratings established by an appropriate laboratory or testing agency.

Spaces containing boilers (main or auxiliary) and oil-fired equipment

The Class rules will state the requirements for portable and semi-portable extinguishers which are employed in spaces containing boilers (main or auxiliary). The Class rules will further indicate whether the requirements for boilers (main or auxiliary) are also applicable to other types of oil-fired equipment, such as oil-fired inert gas generators and oil-fired incinerators. Accordingly, where the Class rules provide firefighting requirements for oil-fired boilers, the same requirements are also applicable to other types of oil-fired equipment. These rules will identify the requirements for the number and type of portable and semi-portable extinguishers to be provided in spaces containing oil fuel units. An oil fuel unit is generally defined as the equipment used for the preparation of oil fuel for delivery to an oil-fired boiler (including inert gas generators, incinerators, and waste disposal units) or equipment used for the preparation for delivery of heated or non-heated oil internal combustion engines and includes any oil pressure pumps, filters, and heaters dealing with oil at a pressure of more than 1.8 bar (26 psi). The Class rule requirements regarding portable and semi-portable extinguishers for spaces containing the above types of equipment are discussed below.

Portable fire extinguishers

Class rules require at least two portable foam extinguishers or the equivalent in each of the following locations:

- Each firing space of a boiler room.
- Each firing space for oil fired equipment (inert gas generators, incinerators, etc.).
- Spaces containing all or part of an oil fuel installation.

In accordance with the requirements stated earlier, portable extinguishers are to be provided in each firing space. Accordingly, if there is more than one firing space in the boiler room, then each firing space should be provided with two portable foam extinguishers. However, where the firing stations for two separate boilers share a common firing space, then a single set of portable foam extinguishers would be acceptable. Unless stated otherwise in the Class rules, the portable extinguishers are to be of the foam-type extinguisher or equivalent. Therefore, other types of portable extinguishers may be provided. However, the alternative extinguisher must have an extinguishing capacity equivalent to the portable foam extinguisher when used on an oil fire. Accordingly, portable foam, CO_2, or dry chemical extinguishers with a 'B-II' rating may be considered acceptable. One of the portable fire extinguishers intended for use in the space should be stowed near the entrance to that space, in accordance with SOLAS regulation II-2/10.3.2.2.

Semi-portable fire extinguisher(s)

Class rules also typically require that there is at least one approved foam-type extinguisher of not less than 135 l (36 gal) capacity or the equivalent in each boiler room. This means that the extinguisher should be provided with a hose on a reel that is suitable for reaching any part of the compartment. Accordingly, at least one 'B-V'-rated semi-portable extinguisher should be provided in each of the aforementioned compartments, and more than one 'B-V' portable extinguisher would be required if the hose of the single unit is not sufficient to reach all portions of the compartment. In the case of an installation involving domestic boilers of less than 175 kW (500 hp) on cargo vessels, special consideration may be given to the elimination of the 'B-V' semi-portable extinguisher in accordance with IACS UI SC30 and IMO MSC Circ.1120.

Table 8.2 references to an oil-fired, inert gas generator or oil-fired incinerator are considered the same equipment as an oil-fired boiler.

Table 8.2 Spaces Containing Boilers and Oil-Fired Equipment

Category 'A' machinery spaces	Type of extinguisher					
	Portable foam applicator (1)	Portable foam	Additional portable foam	135 l Foam	45 l foam (2)	Sandbox (3)9
Boiler room						
Oil-fired boilers	1	2N	—	1 (4)	—	N
Oil-fired boilers and oil fuel units	1	2N + 2	—	1 (4)	—	N
Engine room						
Oil fuel units only	1	2	—	—	—	—

Notes: (in accordance with IACS UI SC30)
(1) May be located outside of the entrance to the room.
(2) May be arranged outside of the space concerned for smaller spaces on cargo vessels.
(3) The amount of sand should be at least 0.1 m3 (3.5 ft3). A shovel should be provided. Sandboxes may be substituted by approved portable fire extinguishers.
(4) Not required for such spaces in cargo vessels wherein all boilers contained therein are for domestic services and are less than 175 kW.
N = Number of firing spaces or stations
2N = Two extinguishers to be located in each firing space.
x = Sufficient (minimum two (2) in each space, thus, at least one portable fire extinguisher is within 10 m (33 ft) walking distance from any point).
y = Sufficient to enable foam to be directed onto any part of the fuel and lubricating oil pressure systems, gearing, and other fire hazards.

Category 'A' machinery spaces containing internal combustion machinery

Category 'A' machinery spaces are defined as those spaces and trunks to such spaces that contain:

- Internal combustion machinery used for main propulsion; or
- Internal combustion machinery used for purposes other than main propulsion where such machinery has in the aggregate a total power output of not less than 375 kW (500 hp); or

- Any oil-fired boiler or oil fuel unit, or any oil-fired equipment other than a boiler, such as an inert gas generator, incinerator, and waste disposal units.

The requirements for portable and semi-portable extinguishers in Category 'A' machinery spaces containing internal combustion machinery will be clearly listed in the appropriate Class rules pertaining to that vessel.

Note: Internal combustion machinery includes any form of internal combustion engine, as well as gas turbines.

Portable fire extinguishers

In accordance with the Class rules, two or more portable foam extinguishers or the equivalent (e.g. 'B-II' rated portable extinguishers) should be provided and are to be so located that no point in the space is more than 10 m (33 ft) walking distance from an extinguisher. One of the portable fire extinguishers intended for use in the space should be stowed within the proximity of the main entrance to machinery space in accordance with SOLAS regulation II-2/10.3.2.2.

Semi-portable fire extinguishers

Each such space is also to be provided with at least one approved foam-type fire extinguisher, each of at least 45 l (12 gal) capacity or the equivalent (e.g. 'B-III' rated extinguishers), and, in accordance with the size of the space or compartment, sufficient in number to enable foam or its equivalent to be directed at any part of the fuel and lubricating oil pressure systems, gearing, and other fire hazards.

Combined boiler and internal combustion engine machinery spaces

Frequently, machinery spaces may contain a combination of boilers, internal combustion engines, and oil fuel units. To provide clarification of the Class rule requirements for portable and semi-portable fire extinguishers in such spaces, the following table is provided.

Note: In Table 8.3, references to an oil-fired, inert gas generator or oil-fired incinerator are considered the same equipment as an oil-fired boiler.

Table 8.3 Combined Boiler/Internal Combustion Engine Machinery Spaces

Category 'A' machinery spaces	Type of extinguisher					
	Portable foam applicator (1)	Portable foam	Additional portable foam	135 l foam	45 l foam (2)	Sandbox (3)
Boiler room						
Oil fuel units only	1	2	N/A	—	—	—
Internal combustion machinery	1	X			y	—
Internal combustion machinery and oil fuel units	1	X			y	—
Combined engine/boiler room						
Internal combustion machinery, oil fired boilers, and oil fuel units	1	(2N + 2) or x whichever is greater		1 (4)	y (5)	N

Notes: (In accordance with IACS UI SC30):
(1) May be located outside of the entrance to the room.
(2) May be arranged outside of the space concerned for smaller spaces on cargo vessels.
(3) The amount of sand should be at least 0.1 m3 (3.5 ft3). A shovel should be provided. Sandboxes may be substituted by approved portable fire extinguishers.
(4) Not required for such spaces in cargo vessels wherein all boilers contained therein are for domestic services and are less than 175 kW.
(5) Machinery spaces containing both boilers and internal combustion engines, one of the foam fire extinguishers of at least 45 l capacity or equivalent, may be omitted on the condition that the 135 l extinguisher can protect efficiently and readily the area covered by the 45 l extinguisher.
N = Number of firing spaces or stations
2N = Two extinguishers to be located in each firing space.
x = Enough (minimum two (2) in each space); thus, at least one portable fire extinguisher is within 10 m (33 ft) walking distance from any point.
y = Sufficient to enable foam to be directed onto any part of the fuel and lubricating oil pressure systems, gearing, and other fire hazards.

Spaces containing steam turbines or steam engines

Spaces containing steam turbines or steam engines used either for main propulsion or for other purposes when such machinery has in the aggregate a total power output of not less than 375 kW (500 hp) are to be provided with the following portable and semi-portable extinguishers:

Portable fire extinguishers

Enough portable foam extinguishers or the equivalent (e.g. 'B-II' extinguishers) are to be provided and located so that no point in the space is more than 10 m (33 ft) walking distance from an extinguisher. As a minimum, there are at least two such extinguishers in each such space. One of the portable fire extinguishers intended for use in the space should be stowed in proximity of the main entrance to that space in accordance with SOLAS regulation II-2/10.3.2.2.

Semi-portable fire extinguishers

Approved foam fire extinguishers, each of at least 45 l (12 gal) capacity or equivalent (e.g. 'B-III' extinguisher), must be provided and are to be in sufficient number so as to enable foam or its equivalent to be directed to any part of the pressure lubrication system, onto any part of the casings enclosing pressure-lubricated parts of the turbines, engines, or associated gearing and onto other fire hazards. However, the Class rules do allow the elimination of these semi-portable extinguishers if an approved fixed fire extinguishing system is installed in the space.

Portable and semi-portable fire extinguishers in other machinery spaces

Machinery spaces other than those discussed earlier, in which a fire hazard exists, are to be provided with enough portable fire extinguishers or other means of fire extinction in, or adjacent to, that space. Machinery spaces are defined as all machinery spaces of category 'A' in addition to all other spaces containing propulsion machinery, boilers, oil fuel units, steam and internal combustion engines, generators and major electrical machinery, oil filling stations, refrigerating, stabilising, ventilation and air conditioning machinery, and similar spaces, and trunks to such spaces. The requirements for category 'A' machinery spaces containing internal combustion machinery, as well as spaces containing oil-fired boilers, oil-fired incinerators, oil-fired inert gas generators, oil fuel units, steam turbines, and steam engines have been addressed earlier. However, there are numerous other 'machinery spaces' found aboard vessels, which present serious fire risks. In accordance with the Class rules, these other machinery spaces must be provided with appropriate portable extinguishers. The following are a few of these spaces:

- Spaces containing internal combustion machinery of less than 375 kW (500 hp) and not used for main propulsion.
- Switchboard and electrical distribution rooms.
- Heating, ventilation, and air conditioning (HVAC) equipment rooms.

- Dedicated pump rooms with electric motors, motor controllers, etc.
- Spaces containing hydraulic power units and hydraulic equipment.
- Oil filling stations.
- Oil fuel units operating at a pressure less than 1.8 bar (26 psi).

It is important to carefully review the vessel's fire control plan to understand the equipment or systems which may be located within various compartments and evaluate the particular fire risk involved to confirm that the necessary portable extinguishers are being provided and are actually suitable for the service. The above only identifies a few of the 'other machinery spaces' found aboard a vessel which pose a fire risk. Any 'other machinery space' which poses a fire risk must be provided with suitable portable extinguishers.

Accommodation, service spaces, and control stations

The vessel's accommodation, service spaces, and control stations must be provided with portable fire extinguishers suitable for their intended service. Despite this being a very generic statement, it has very far-reaching effects. To properly comply with this requirement, we must (1) clearly understand which spaces are included under the rather generic term 'accommodation, service spaces and control stations' and (2) understand the fire risks typically inherent with the various spaces to confirm that the portable extinguishers being provided are suitable for the intended service. For the first item noted earlier, the Class rules provide very specific definitions for these spaces and the same are provided below:

(1) Accommodation spaces.
     Accommodation spaces are those spaces used for public spaces, corridors, lavatories, cabins, offices, hospitals, cinemas, games and hobbies rooms, barber shops, pantries containing no cooking appliances, and similar spaces. Furthermore, public spaces are those portions of the accommodation which are used for halls, dining rooms, lounges, and similar permanently enclosed spaces.
(2) Service spaces.
     Service spaces are those spaces used for galleys, pantries containing cooking appliances, lockers, mail and specie rooms, storerooms, workshops other than those forming part of the machinery spaces, and similar spaces and trunks to such spaces.
(3) Control stations.
     Control stations are those spaces in which the vessel's radio, main navigation equipment, or emergency source of power is located or where the fire recording or fire control equipment is centralised:
     - Spaces containing, for instance, the following battery sources are to be regarded as control stations regardless of battery capacity.

- Emergency batteries in separate battery room for power supply from blackout till start of emergency generator.
- Emergency batteries in separate battery room as reserve source of energy to radio telegraph installation,
- Batteries for the start of emergency generator(s).
- In general, all emergency batteries.

For the second part of this requirement, it is important to consider the specific fire hazards that might be involved with the specific location. In most accommodation spaces such as cabins, offices, and public spaces, the greatest concern would be class 'A' fires. However, in service spaces, the concern may more likely be around class 'B' fires (liquid fuel fires) or even class 'C' fires (electrical fires). As stated previously, class 'F' or 'K' fire is only prevalent in the galley and food preparation areas.

Accommodation spaces

The Class rules require that all corridors are provided with suitable portable fire extinguishers located no more than 45 m (150 ft) apart. For normal accommodation spaces (for instance, cabins, offices, games and hobbies rooms, barbers shops, pantries containing no cooking appliances and similar accommodation spaces), type 'A-II' portable extinguishers provided at the above spacing on each level are usually considered acceptable. Typically, additional 'A-II' portable extinguishers are provided in larger public spaces, such as messes, dining rooms, and lounges. However, there are certain other accommodation spaces such as hospitals and cinemas that may warrant additional types of extinguishers due the fire hazards involved.

Service spaces

As noted from the definition, the term 'service spaces' may include a wide variety of spaces and a variety of fire hazards. The installation of type 'B-II' portable extinguishers would be appropriate for galleys and possibly workshops, while type 'A-II' should be provided for mail, specie, and storerooms. Once again, it is very important to account for the fire hazard involved when considering the suitability of the type of portable extinguisher being provided.

Control stations

Fire hazards associated with control stations mainly involve electrical equipment. Accordingly, type 'C-II' extinguishers should normally be provided for these spaces. For vessels of 1,000 GRT and above, the total number of portable extinguishers for these spaces should never be less than five. In addition, the number of portable extinguishers provided for a particular space should be based on the amount and type of equipment or combustible materials within that space. Furthermore, where portable extinguisher(s) is provided to protect a specific space (i.e. control station and galley), one of the portable fire extinguishers should be stowed near the entrance to that space. As discussed earlier, carbon dioxide fire extinguishers should not be placed in accommodation spaces as their use can lead to asphyxiation. In control stations and other spaces containing electrical or electronic equipment or appliances necessary for the safety of the vessel, fire extinguishers should be provided whose extinguishing media are neither electrically conductive nor harmful to the equipment and appliances. As per SOLAS regulation II-2/10.3.2.4, fire extinguishers should be situated ready for use at easily visible places, which can be reached quickly and easily at any time in the event of a fire. Furthermore, they should be positioned in such a way that their serviceability is not impaired by the weather, vibration, or other external factors. Portable fire extinguishers should be provided with devices that indicate whether they have been previously used.

Fire protection arrangements for paint lockers

Paint lockers and flammable liquid lockers can contain large quantities of paints, cleaners, solvents, and other toxic products that may have low flash points. While these products are stored in sealed containers, they can represent a sizable fire load. In addition, these areas are normally also used for mixing operations, resulting in flammable atmospheres. This creates a very real potential for fires. To address the hazards associated with paint and flammable liquid lockers and to fight any fires that may develop, approved fire extinguishing arrangements are needed. Normally, paint lockers and flammable liquid lockers are required to be provided with a fixed fire extinguishing system. However, for paint lockers and flammable liquid lockers with a deck area less than 4 m^2 (43 ft^2), which provide no access to accommodation spaces, portable 5 kg (11 lbs) CO_2 extinguishers, which can be discharged through a port in the boundary of the lockers, may be acceptable instead. The number of portable CO_2 extinguishers to be provided should be sufficient to provide 40% of the gross volume of the compartment when considering the specific volume of CO_2 to be 0.56 m^3/kg (9 ft^3/lb).

Fire protection arrangements for helicopter landing areas

Where the vessel is fitted with arrangements to accommodate helicopter operations, the following arrangements must be provided in accordance with Class rules.

Semi-portable extinguishers

The helicopter deck must be protected with a minimum of two approved dry powder extinguishers, with a total capacity of not less than 45 kg (100 lbs).

CO_2 extinguishers

The helicopter deck should be protected by CO_2 extinguishers with a total capacity of not less than 18 kg (40 lbs) or equivalent. At least one of these extinguishers must be equipped so as to enable it to reach the engine area of any helicopter using the helicopter deck. The CO_2 extinguishers are to be located so that the equipment would not be vulnerable to the same damage as the dry powder extinguishers.

VESSELS WITH UNMANNED MACHINERY SPACES (UMS)

Vessels with unmanned machinery spaces (UMS), and which are classed with the 'ACCU' automation notation or similar, are required to have a dedicated 'Firefighting Station'. Moreover, certain additional equipment is required to be provided at the 'Firefighting Station'. This additional equipment includes several portable fire extinguishers equal to the complement of machinery space portable fire extinguishers. However, where duplicated portable extinguishers are provided to satisfy Class requirements in lieu of spare charges, these duplicated extinguishers may be considered to satisfy the earlier requirement, provided they are stored at the 'Firefighting Station'.

ADDITIONAL REQUIREMENTS

RORO spaces, RORO spaces carrying motor vehicles with fuel in their tanks, and cargo spaces carrying motor vehicles with fuel in their tanks (other than RORO spaces)

In accordance with Class rules, at least one portable extinguisher should be provided at each access to any RORO cargo space. In addition, at each vehicle deck level where vehicles with fuel in their tanks are carried, sufficient portable extinguishers suitable for fighting oil fires (B-II extinguishers) are to be provided such that they are not more than 20 m (65 ft) apart on both sides of the vessel. The one located at the access may be credited for this purpose. It should be noted that when applying the requirements of SOLAS regulation II-2/20.6.2.1, IMO MSC.1/Circ.1275 provides an interpretation, which specifies that the portable fire extinguishers being provided in the above-listed spaces are suitable for class 'B' fires and the number and capacity of the extinguishers are arranged as follows:

- Two portable fire extinguishers, each having a capacity of not less than 6 kg (13.2 lbs) of dry powder or equivalent, should be provided when dangerous goods are carried on the weather deck, in open RORO spaces and vehicle spaces, and in cargo spaces as appropriate. Two portable fire extinguishers, each having a suitable capacity, should be provided on weather deck for tankers.
- No portable fire extinguisher needs to be provided in cargo holds of container ships if motor vehicles with fuel in their tank for their own propulsion are carried in open or closed containers.

Vessels carrying IMDG classed dangerous goods

If 'Dangerous Goods' cargoes are to be carried onboard the vessel, additional portable extinguishers may be required. For vessels carrying Dangerous Goods with a Class 3 designation, and a flash point less than 23 °C (73 °F) or Class 3 with a flashpoint equal to or more than 23 °C (73 °F) but not more than 60 °C (140 °F), Classes 4.1, 4.2, 4.3 liquids and 4.3 solids; Class 5.1, Class 6.1 liquids with a flashpoint less than 23 °C (73 °F) or Class 6.1 liquids with a flashpoint equal to or more than 23 °C (73 °F) but not more than 60 °C (140 °F), and Class 8 liquids with a flash point less than 23 °C (73 °F) or Class 8 liquids with a flash point equal to or more than 23 °C (73 °F) but not more than 60 °C (140 °F) are required to be provided additional portable extinguishers. These additional extinguishers are to be dry powder portable extinguishers with a total capacity of at least 12 kg (26.5 lbs) or the equivalent.

Chemical tankers

Chemical carriers must be provided with suitable portable fire extinguishing equipment for the products being carried.

Vessels under 90 m (295 ft)

For vessels under 90 m (295 ft) in length, refer to the specific Class rules regarding guidance on the type, number, and size of portable extinguishers required to be provided onboard. For an indication, the ABS Class rules stipulate the following number of portable extinguisher on vessels under 90 m (295 ft) in length:

Table 8.4 Spaces Containing Boilers and Oil-Fired Equipment

Space	Classification	Quantity and location [5]
Safety areas		
Communicating corridors	A-II	One in each main corridor not more than 46 m (150 ft) apart (may be located in stairways).
Pilothouse/bridge	C-II	Two in vicinity of exit. See notes 4 and 6.
Radio room	C-II	One in vicinity of exit. See note 4.
Accommodation		
Sleeping accommodation	A-II	One in each sleeping accommodation space (where occupied by more than four persons).
Service spaces		
Galleys	B-II or C-II	One for each 230 m² (2,500 ft²) or fraction thereof for hazards involved.
Storerooms	A-II	One for each 230 m² (2,500 ft²) or fraction thereof located in vicinity of exits, either inside or outside spaces. See note 4.
Workshops	A-II	One outside the space in vicinity of exit. See note 4.
Machinery spaces		
Internal combustion or gas turbine engines	B-II and	One for each 746 kW (1,000 hp), but not less than 2 nor more than 6. See note 1.
	B-III	One required. See note 3.
Electric motors or generators of the open type		One for each motor or generator unit. See note 2.

Notes:
1 Where the installation is on the weather deck, or open to atmosphere, one B-II for every three engines is allowable.
2 Small electrical appliances such as fans are not counted or used as the basis for determining the number of extinguishers required.
3 Not required for vessels of less than 500 GRT.
4 Vicinity is intended to mean within 1 m (3 ft).
5 For vessels of 1,000 GRT and above, at least five extinguishers are to be provided for accommodation spaces, services spaces, or spaces where the vessel's radio, main navigation equipment, or emergency source of power is located, and locations where the fire recording or fire control equipment is located.
6 For cargo ships less than 500 GRT, 'C-I' portable extinguishers may be used.

Oil tankers under 30.5 m (100 ft)

Oil tankers with a length less than 30.5 m (100 ft) may be provided with two 'B-V' extinguishers for the protection of the cargo tank area in lieu of a deck foam system.

Spare charges

A spare charge should be provided for 100% for the first ten extinguishers and 50% for the remaining extinguishers, but no more than 60 total spare charges are required. For fire extinguishers, which cannot be recharged by the crew, additional portable fire extinguishers of the same quantity, type, capacity, and number should be provided in lieu of spare charges. Instructions for recharging should be carried onboard. Only refills approved for the fire extinguisher in question may be used for recharging. Partially emptied extinguishers should also be recharged.

In Part 2 of this book, we have covered a lot of ground relating to the various systems and infrastructure used on shipboard settings to fight fires. We have also examined the universally accepted Class rules and requirements relating to these systems. In Part 3, we will turn our attention to the additional requirements, which, although intrinsic to shipboard firefighting, largely fall outside the main scope of Class rules, and are therefore considered 'additional requirements'.

Part 3

Additional requirements

Chapter 9

Additional fire protection requirements

On a vessel, the proximity of fuel sources and numerous sources of ignition present special concerns for fire safety. This chapter discusses certain Class rule requirements that are intended to address these concerns. However, as always, reference should be made to the Class rules applicable to the specific vessel concerned for the most current applicable Class rules.

SEGREGATION OF FUEL OIL PURIFIERS FOR HEATED OIL

Any equipment intended to handle or process oil constitutes a significant fire hazard, especially where heated oil is involved. The vapours from fuel leaks or spilled oil at elevated temperatures need only contact a source of ignition to initiate a serious fire. One piece of equipment providing such concerns is the fuel oil purifier unit. To minimise the potential of oil escaping from a purifier and coming into contact with one of the various sources of ignition within the engine room, fuel oil purifiers containing heated oil are required to be installed in a separate compartment or space. Such fuel oil purifier rooms are to be enclosed with steel bulkheads extending from deck to deck and provided with self-closing doors. In addition, Class rules require these rooms to be provided with the following:

- Independent mechanical ventilation or ventilation arrangements that can be isolated from the machinery space ventilation. These must be of the suction type.
- A fire detection system.
- A fixed fire extinguishing system capable of activation from outside the room. The extinguishing system should be dedicated to the fuel oil purifier room but may form part of the fixed fire extinguishing system for the entire machinery space. However, for the protection of purifiers on cargo vessels of 2,000 GRT and above located within a machinery space of category 'A' above 500 m^3 (17,657 ft^3) in volume, the aforementioned referenced fixed dedicated system must be a fixed

water-based system or equivalent, or alternatively a local application fire extinguishing system. The system should be capable of activation from outside the fuel oil purifier room. In addition, protection should be provided by the fixed fire extinguishing system covering the Category 'A' machinery space in which the purifier room is located.
• Means of closing any ventilation openings, and stopping the ventilation fans, purifiers, purifier-feed pumps, etc., from a position close to where the fire extinguishing system is activated.

The Class rules incorporate a provision for 'special consideration' where it can be established that it is 'impracticable' to locate fuel oil purifiers in a separate compartment or space. However, any special consideration must establish that it is truly 'impracticable' to locate this equipment in a separate compartment or space, and that it is not just a matter of cost or convenience. Decisions to grant any allowances must be based on the fuel oil purifier location, the containment of possible leaks, shielding and ventilation, in conjunction with the specific size of the vessel and of the engine room, and documentation that must establish that it is truly not practicable to incorporate such arrangements. The application of this 'special consideration' would appear to be applicable to only small vessels as the engine rooms of larger vessels should be designed to incorporate the necessary segregation. Therefore, if it is impracticable to locate the fuel oil purifiers in a separate compartment or space, special consideration may be given with respect to the location, containment of possible leaks, shielding, and ventilation. In such cases, a local fixed water-based fire extinguishing system may be provided instead. Where, due to the limited size of the category 'A' machinery space (i.e. less than 500 m3 (17,657 ft3) in volume), a local fixed water-based fire extinguishing system is not required to be provided, then an alternative type of local dedicated fixed fire extinguishing system must be provided for the protection of the fuel oil purifiers. In either instance, the local fire extinguishing system should activate automatically or manually from the centralised control station or some other suitable location. If an automatic release is provided, an additional manual release is also to be arranged. Compliance with the above requirement should be verified during the review of the machinery room arrangement drawing.

SEGREGATION OF HIGH-PRESSURE HYDRAULIC UNITS

Hydraulic units with a working pressure above 15.5 bar (225 psi) located within a machinery space are required to be placed in a separate compartment or space. Any piping outside of this compartment or space should have as few joints as are practicable, and those joints should be shielded, as necessary, to prevent any oil or oil mist that may escape under pressure from contacting hot surfaces, open electrical equipment that could ignite the oil, or any other sources of ignition. Compliance with the segregation requirement should be verified during the review of a hydraulic system by consulting the machinery room arrangement drawing or other suitable arrangement drawing.

PIPING SYSTEMS HANDLING OIL

Material requirements for piping systems conveying oil

To eliminate the potential of adding additional fuel to a fire, all fuel, lube oil and hydraulic piping, valves, and fittings must be constructed of steel or some other approved material, in accordance with SOLAS regulation II-2/4.2.2.5.1, II-2/4.2.3.1, and II-2/4.2.4. By requiring piping components of systems conveying oil to be manufactured from steel or some other approved materials, the piping system incorporates sufficient structural integrity to resist physical abuse or damage. Furthermore, steel with a solidus melting point above 927 °C (1,700 °F) is not readily affected by heat. It is recognised that there is a need for flexibility at certain points within the fuel and lube oil systems (i.e. connection to the engines), and therefore, the use of non-metallic hoses is permitted. However, these hoses are required to be fire-resistant and pass Class-mandated flame test criteria.

Remote closure arrangements for the valves on oil tanks

Fuel oil and lube oil utilised for the ship services and propulsion are stored in large capacities and typically located in the immediate vicinity of the engine room. The capacity of these tanks can be as high as 3,785 m³ (1,000,000 gal) depending on the size of the vessel. In the event of a fire, a means is needed to prevent the contents of the tank from adding fuel to the fire, even if the associated piping is damaged. Accordingly, Class rules require that all pipes which enter a storage, settling or daily service lube or fuel oil tank, and at such a level that they could be subjected to a static head of oil from the tank, must be fitted with positive closing valves at the tank which are capable of being closed from outside the space concerned in the event of a fire. Furthermore, these tank valves are required to be of steel or equivalent material and must have an elongation of at least 12% if installed outside the tank.

Fuel oil and lube oil systems on engines

The operation of propulsion and generator engines requires fuel oil and lube oil systems utilised at high pressures and temperatures and clearly in the direct vicinity of heated surfaces and other sources of ignition. Accordingly, there is the potential for serious fires if these piping systems should leak or release oil in any way. To address these concerns, Class will set certain rules and requirements which are specific to the type and size of vessel. These typically include special limitations on the type of piping that can be used, the arrangements to prevent oil from being released when strainers are inspected or cleaned, as well as leak containment and other ancillary arrangements required to protect against oil coming into contact with sources of ignition.

Insulation requirements for heated surfaces

The surfaces of internal combustion engines, gas turbines, boilers, and exhaust pipes can often reach temperatures high enough that any flammable liquid, which may come into proximity with these surfaces, would cause a major fire hazard. Therefore, it is a requirement that any surface that may be heated to a temperature in excess 220 °C (428 °F), and may be exposed to a flammable liquid, be insulated with some form of approved non-combustible material(s). If this insulation is oil absorbing and exposed to penetration of oil, then the insulation must be encased in sheet metal or some other form of insulation that is not capable of being oil penetrated.

Remote stopping of fuel oil pumps and thermal fluid circulating pumps

To minimise the spread of a fire, it is important to eliminate any additional sources of fuel that could feed the fire. Accordingly, arrangements are required to shut off auxiliary machinery, such as oil-fuel transfer pumps, oil-fuel unit pumps, and other similar fuel pumps and thermal-oil heating pumps. The shut-off arrangements for this type of equipment are required to be located outside the space concerned so that they may be stopped in the event of a fire arising in the space where they are located and include the following:

- Oil-fuel pump remote stops.
- Remote stops for circulating pumps for thermal oil heating systems.
- Purifier room remote closures and stops.

Remote stops for ventilation fans and closing arrangements for openings

Like the need to stop fuel oil and thermal heating pumps, since oxygen is vital to the propagation of fire, any fans supplying air to a space engaged in a fire should be stopped immediately. This will minimise the supply of oxygen to the fire and assist in bringing the fire under control. Accordingly, all ventilation fans are to have a remote means of stopping. The remote stop should be located outside the space(s) being ventilated by the fans with arrangements to close the ventilation inlets and outlets also required. In particular, the remote means of stopping the ventilation fans serving the machinery spaces are to be situated in the passageway leading to the engine room or at the fire control station. In addition, it is important that any compartment involved in a fire can be made as airtight as possible, to minimise the inflow of oxygen into the involved compartment. This is especially important for spaces protected by a fixed gas fire extinguishing system. Accordingly, means for closing all doorways, ventilators, annular spaces around funnels, and other openings to machinery rooms, pump rooms, and cargo spaces must be provided.

PAINT AND FLAMMABLE LIQUID LOCKERS

Most paints, varnishes, lacquers, and enamels, as well as numerous types of cleaners that are used onboard are oil-based, and therefore, present a high fire risk. Accordingly, these flammable liquids must be stowed in one or more designated paint or flammable liquid lockers or rooms. Due to the high risk of fire, paint lockers and similar service spaces used for the storage of flammable liquids are required to be protected by an appropriate fire extinguishing arrangement, which is further discussed below.

Lockers of 4 m^2 (43 ft^2) and more floor area

Paint lockers and flammable liquid lockers with a floor area of 4 m^2 (43 ft^2) or more and also such lockers of any floor area with access to accommodation spaces are to be provided with one of the fixed fire extinguishing systems specified below:

- CO_2 system, designed for 40% of the gross volume of the space.
- Dry powder system, designed for at least 0.5 kg/m^3 (0.03 lb/ft^3).
- Water spraying system, designed for 5 l/min/m^2 (0.12 gpm/ft^2). The water spraying system is permitted to be connected to the vessel's fire main system, in which case the fire pump capacity should be sufficient for simultaneous operation of the fire main system and the water spray system. Precautions are also required to be taken to prevent the nozzles from becoming clogged by impurities in the water or corrosion of piping, nozzles, valves, and pump.
- Systems other than those mentioned earlier may be considered, provided they are not less effective.

Lockers of less than 4 m² (43 ft²) floor area

For paint lockers and flammable liquid lockers of floor area less than 4 m² (43 ft²) having no access to accommodation spaces, portable fire extinguisher(s) sized in accordance with the Class rules, and which can be discharged through a port in the boundary of the lockers, may be accepted instead. The required portable fire extinguishers are to be stowed adjacent to the port. Alternatively, a port or hose connection may be provided for this purpose to facilitate the use of water from the fire main.

Chapter 10

Fire control plans

The fire control plan (FCP) provides vital information that is crucial for the rapid and efficient action of the vessel's crew during a fire. Accordingly, it is very important that the FCP accurately reflects the firefighting arrangements installed onboard and is consistent with the arrangements approved by Class for the vessel. The following discussion is limited to the 'active' firefighting arrangements and equipment for the vessel to be depicted on the FCP. This discussion does not address requirements associated with passive fire protection, such as structural fire protection, or any additional requirements associated with electrical systems or arrangements. However, other requirements for 'passive' fire protection arrangements (e.g., structural fire protection arrangements and escape routes), as well as electrical system requirements (e.g., electrical control arrangements and emergency generator location), exist and **should** be shown on the FCP. The FCP is frequently submitted as part of the Ship's Safety Plan (SSP), which in turn is an integral component of the Ship's Safety Management System (SMS), which also indicates the quantity and position of lifesaving appliances and equipment, communications equipment, etc. Whereas most Class administrations are authorised to review the requirements associated with fire control and safety plans by most Flag States, the review for lifesaving equipment falls outside the scope of this book.

STANDARDISED SYMBOLS

The IMO has issued IMO Resolution A.952(23), '*Graphical symbols for shipboard fire control plans*' which provides a standardised set of symbols to be used on the vessel's FCP. The use of a standardised set of symbols is intended to assist the vessel's officers and crew in quickly identifying and locating equipment, and is very helpful in making the FCP 'user friendly'. It is worth noting, however, the resolution was issued as guidance only and its use is not mandated by SOLAS. Accordingly, whilst there are certain Flag States that require the use of the standardised symbols identified in IMO

Res. A.952(23), its use is not mandatory insofar as Class requirements are concerned.

Related IMO MSC 1120 Interpretation.
SOLAS Reg. II-2/20 (Consolidated Edition: 2014) II-2/15).
Reference to Resolution A.952(23).

Reference is made to resolution A.952(23) – Graphical symbols for fire control plans and resolution A.765(18) – Guidelines on the information to be provided with fire control plans and booklets required by SOLAS regulations II-2/20 (SOLAS (Consolidated Edition 2014) II-2/15).

CLASS REQUIREMENTS FOR FCPs

The following information is provided as general guidance on the Class requirements for FCPs. However, reference should always be made to the Class rules applicable to the specific vessel concerned for the complete set of requirements.

Steel vessels of 90 m in length and greater in unrestricted service

Information and details

FCPs are to be general arrangement plans that show the following details for each deck, as applicable:

- The location of control stations.
- Various fire sections enclosed by class 'A' divisions.
- Sections enclosed by class 'B' divisions.
- The particulars of the fire detection and alarm systems.
- The particulars of the sprinkler installation.
- The particulars of the fire extinguishing appliances.
- Any means of access to different compartments, spaces, decks, etc.
- The ventilation system including the particulars of the fan control positions, the position of dampers, and the identification numbers of the ventilating fans serving each section.

As an alternative to using general arrangement plans, the details listed earlier may instead be set out in a booklet. If this approach is selected, a copy of the booklet is required to be supplied to each of the ship's officers, and one copy must be kept available onboard, in an accessible position at all times. Moreover, the FCP (and/or booklet) must be kept up to date, with any alterations being recorded thereafter as soon as practicable. The descriptions in the FCP (and/or booklet) must be in the official language of the vessel's Flag State. Where that language is neither English nor French, a translation into either

English or French is required. In addition, instructions concerning the maintenance and operation of all the equipment and installations onboard for the fighting and containment of fire are to be kept under one cover, and readily available in an accessible location. A copy of the FCP (and/or booklet) must be permanently stored in a prominently marked weathertight enclosure outside the bridge for the assistance of shoreside firefighting personnel in accordance with the provisions set out in IMO MSC/Circ.451 – *Guidance concerning the location of Fire Control Plans for assistance of shore side firefighting personnel*. From the aforementioned list of items to be shown on the FCP, the details and arrangements that the Class engineering services department are likely to review include any or all of the following:

- Number and location of control stations.
- The particulars of the fire detection and alarm systems.
- The particulars of the sprinkler installation.
- The particulars of the fire extinguishing appliances.
- The ventilation system including particulars of the fan control positions and identification numbers of the ventilating fans serving each section.

Control Stations

Control stations as those spaces in which the vessel's radio or main navigating equipment or the emergency source of power is located or where the fire recording or fire control equipment is centralised. In addition, spaces containing the following battery sources are to be regarded as control stations, regardless of the battery capacity:

- Emergency batteries located in separate battery rooms for power supply from blackout till the start of the emergency generator.
- Emergency batteries located in separate battery rooms forming a reserve source of energy to the vessel's radio telegraph installation.
- Batteries for starting the emergency generator.
- In general, all emergency batteries required in accordance with Class requirements.

The review of the FCP should verify that the above locations are clearly identified.

Fire detection and fire alarm systems

Where a fire detection and alarm system are required and/or installed, the locations of the control panels, indicating units, manual call points, and detectors must be clearly indicated on the FCP. Compliance with Class requirements, and consistency with the arrangements indicated on the approved fire detection and alarm system diagram, should be verified during the Class review of the FCP.

Sample smoke detection and alarm systems

Where a smoke detection and alarm system are required and/or installed, the locations of the control panels, indicating units, and detectors must be clearly indicated on the FCP. Compliance with Class requirements, and consistency with the arrangements indicated on the approved smoke detection and alarm system diagram, should be verified during the Class review of the FCP.

The particulars of sprinkler installations and fire extinguishing appliances

A substantial number of the fire extinguishing systems and appliances found on shipboard settings include, as a minimum, the following:

- Fire main system arrangements.
- Fixed fire extinguishing system locations and arrangements.
- Sprinkler system locations and arrangements.
- Provision, location, and particulars of portable firefighting equipment and appliances.
- Location of controls for fuel-oil pumps and oil tank shut off valves.
- Provision, location, and particulars of firefighter's outfits.
- The number and locations of Emergency Escape Breathing Devices (EEBDs).

Details should be indicated on the FCP for each of the above systems or equipment and further discussion regarding each of these items is provided below.

Fire main system

The number and location of the main fire pumps must be clearly indicated on the FCP. The arrangements are to indicate that the location(s), as well as the separation and access arrangements, comply with the Class requirements for cargo vessels. The location and access arrangements for the emergency fire pump, if required, must be clearly indicated on the FCP and comply with Class requirements. Whilst it may be not specifically required by Class, the identification of those spaces where the fire pump isolation valves required by Class are located is useful to the vessel's officers and crew. The review should verify that the locations are indicated and are consistent with the locations indicated on the approved fire main system diagram. The number and position of the hydrants must comply with Class requirements and be consistent with the arrangements indicated on the approved fire main system diagram. The locations of the hydrants should be reviewed, and any questionable areas brought to the attention of the submitter and the Class Surveyor. Frequently, the FCP will not indicate the maximum coverage areas from each hydrant, and therefore, accurate verification of compliance is difficult from the drawing alone. Also, actual arrangements onboard may involve certain restrictions, which are not readily apparent from the drawing. Accordingly, the attending Class Surveyor should also be requested to verify compliance with the Class rule requirements. In most cases, this requires that at least one fire hose is provided, each complete with couplings and nozzles, for each 30 m (100 ft) length of the vessel, as well as one additional hose with couplings and nozzles, as a spare. Furthermore, for vessels of 1,000 GRT and above, a minimum of five hoses are required. The above does not include any hoses required in any engine space or compartment, or for the boiler room. The FCP should indicate compliance with the above requirement. The FCP should indicate the individual hose lengths, and the review should verify that the lengths are in accordance with Class requirements. The FCP should indicate the type of fire hose nozzles, and the review should verify that they are of the dual-purpose type and comply with Class requirements. At least one International Shore Connection (ISC) is to be shown on the FCP, wherein the FCP should indicate at least two locations where the ISC can be attached to the fire main system, that is, on either side of the vessel.

Fixed fire extinguishing systems

The following spaces are required to be fitted with a fixed fire extinguishing system:

- All main or auxiliary boiler rooms.
- All spaces containing internal combustion engines or gas turbines with an output greater than 375 kW (500 hp) (i.e. category 'A' machinery space).
- All spaces containing steam turbines or steam engines with an output greater than 375 kW (500 hp) and intended for periodic unattended operation (UMS).
- Any space containing oil fuel units, oil-fired inert gas generators, and oil-fired incinerators.
- Paint or flammable liquid lockers, depending on their size.
- The accommodation spaces of cargo vessels, depending on the method of construction.
- Any cargo spaces as specifically required by Class.

The Fire Control Plan should be reviewed to verify that the fixed fire extinguishing systems are indicated for these spaces. In addition, the location of the system release controls, as well as storage arrangements for the fire extinguishing medium, should be properly verified.

Portable extinguishers

The type, size, and locations of all portable extinguishers must be indicated on the FCP and comply with the Class rules. This will be verified during Class review of the plan. However, as a detailed discussion concerning the requirements regarding the types, designs, and locations of portable firefighting extinguishers is provided in Chapter 8, the requirements are not repeated here.

Ventilation fans

The locations and identification numbers for the ventilation fans serving the various spaces should be clearly indicated on the FCP.

Firefighter's outfits

Details regarding the firefighter's outfits are required to be clearly indicated on the FCP. The location and type of equipment to be provided as part of the firefighter's outfits are addressed, in summary, below. The following provides a brief discussion of the required equipment and the locations of the firefighter's outfits onboard shipboard settings:

- *Number of firefighter's outfits.* At least two firefighter's outfits consisting of the equipment identified below are to be carried onboard each vessel. However, certain types of vessels require additional sets of firefighter's outfits, depending upon the type of vessel involved. Refer to the Class rules for specific vessel types (for instance, oil tankers and gas carriers).
- *Individual firefighter's outfit.* Each firefighter's outfit should consist of a set of personal protective equipment and breathing apparatus, to include the following:
  o *Personal protective equipment.* Personal protective equipment should consist of:
  o Protective clothing of an approved material to protect the skin from the heat radiating from the fire and from burns and scalding by steam. The outer surface should be water-resistant.
  o Boots made of rubber or some other form of approved electrically non-conducting material.

o A rigid helmet providing effective protection against impacts from above.

o An electric safety lamp (hand lantern) of an approved type with a minimum burning period of three hours. Electric safety lamps on tankers and those intended to be used in hazardous areas must be of an intrinsically safe, explosion proof type.

o An axe with a handle provided with high-voltage insulation.

- *Breathing apparatus.* Breathing apparatus must be a self-contained compressed air-operated type. The volume of the air contained in the breathing apparatus cylinders must either:

o Hold a minimum of 1,200 l (317 gal), or

o Be of some other self-contained breathing apparatus, which should to be capable of functioning for at least 30 minutes.

Two spare charges must be provided for each required breathing apparatus. All air cylinders for breathing apparatus are to be inter-changeable. In accordance with the FSS Code, Chapter 3/2.2.2, com-pressed air breathing apparatus must be fitted with an audible and visual alarm or some other approved device which will alert the wearer before the volume of air in the cylinder has been reduced to no less than 200 l (52 gal).

- *Lifeline.* For each breathing apparatus, a fireproof lifeline of at least 30 m (98.5 ft) in length must be provided. The lifeline should success-fully pass an approval test by static load of 3.5 kN (787 lbf) for five minutes without failure. The lifeline should be capable of attaching by means of a snap hook to the harness of the apparatus or to a separate belt to prevent the breathing apparatus from becoming detached when the lifeline is operated.

- *Locations of firefighter's outfits.* The firefighter's outfits and equip-ment are required to be stored so to be easily accessible and ready for use and are to be stored in widely separate positions. The locations for the storage of firefighter's outfits and personal protective equipment should be permanently and clearly marked.

- *Firefighter's communication.* For ships constructed on or after 1 July 2014, a minimum of two two-way portable radiotelephone apparatus for each fire party for firefighter's communication must be carried onboard. These two-way portable radiotele-phone apparatus must be of an approved intrinsically safe, explosion proof type. Ships constructed before 1 July 2014 were required to comply with this requirement no later than the first survey after 1 July 2018 (in accordance with SOLAS regulation II-2/10.10.4).

Emergency escape breathing devices

To aid the crew in escaping from smoke filled spaces, vessels must be provided with EEBDs. The standardised requirements for the carriage and use of EEBDs are outlined below:

- *Accommodation spaces.* All vessels are required to carry a minimum of two EEBDs, with at least one spare device within the accommodation spaces.
- *Machinery spaces.* On all vessels, within the machinery spaces, EEBDs are to be situated ready for use at clearly visible locations, which can be reached quickly and easily at any time in the event of fire. The location of EEBDs is to account for the layout of the machinery space and the number of personnel normally required to work in such spaces. Further guidance is provided in MSC/Circ. 849 and 1081: the *Guidelines for the performance, location, use and care of emergency escape breathing devices.* The number and locations of EEBDs are to be indicated in the FCP.

A summary of the MSC/Cir.1081 requirements is shown in Table 10.1. This applies to machinery spaces where crew members are normally employed or are present on a routine basis (i.e. non-UMS).

Table 10.1 Minimum Number of Required EEBDs

A. In machinery spaces for Category 'A' containing internal combustion machinery used for main propulsion [1]:

a) 1 × emergency escape breathing device in the engine control room, if located within the machinery space

b) 1 × emergency escape breathing device in workshop areas. If there is, however, a direct access to an escape way from the workshop, an emergency escape breathing devices is not required; and

c) 1 × emergency escape breathing device on each deck or platform level near the escape ladder constituting the second means of escape from the machinery space (the other means being an enclosed escape trunk or watertight door at the lower level of the space).

B. In machinery spaces of Category 'A' other than those containing internal combustion machinery used for main propulsion,

1 × emergency escape breathing device should, as a minimum, be provided on each deck or platform level near the escape ladder constituting the second means of escape from the space (the other means being an enclosed escape trunk or watertight door at the lower level of the space).

C. In other machinery spaces

The number and location of emergency escape breathing devices are determined by the Flag State administration.

Note:
(1) Alternatively, a different number or location may be determined by the Flag State administration taking into consideration the layout and dimensions or the normal manning of the space.

- EEBD specification:
  - o General. An EEBD is a supply air or oxygen device only used for escape from a compartment that has a hazardous atmosphere and should be of an approved type. EEBDs are not approved to be used for fighting fires, entering oxygen deficient voids or tanks, or worn by firefighters. In these events, a self-contained breathing apparatus, which is specifically suited for such applications, should be used.
  - o Particulars. The EEBD should have a duration of service of ten minutes. The EEBD should include a hood or full-face piece, as appropriate, to protect the eyes, nose, and mouth during escape. Hoods and face pieces are to be constructed of flame-resistant materials and include a clear window for viewing. An inactivated EEBD should be capable of being carried hands-free.
  - o Storage. An EEBD, when stored, should be suitably protected from environment.
  - o Instructions and markings. Brief instructions or diagrams clearly illustrating their use must be clearly printed on the EEBD. The donning procedures are to be quick and easy to allow for situations where there is little time to seek safety from a hazardous atmosphere. Maintenance requirements, manufacturer's trademarks and serial number, shelf life with accompanying manufacture date, and name of approving authority are to be printed on each EEBD. All EEBD training units must be clearly marked.

Fire protection arrangements for helicopter operations

If the vessel is fitted with a helicopter deck arranged to accommodate helicopter operations, the following firefighting arrangements are to be provided, in accordance with Class rules, and the same should be indicated on the FCP. Verification that the appropriate arrangements are indicated should be determined during the review of the FCP.

- *Helicopter decks.* Helicopter decks are defined as a helicopter landing area on a vessel, including all structure, firefighting appliances, and other equipment necessary for the safe operation of helicopters. Helicopter decks are to be fitted with the following firefighting systems and portable extinguishers:
  - o *Hoses and nozzles.* At least two approved combination solid stream and water spray nozzles and hoses sufficient in length to reach any part of the helicopter deck.
  - o *Portable extinguisher.* The helicopter deck should be protected by at least two approved dry powder extinguishers with a total capacity of not less than 45 kg (100 lbs).
  - o *Back-up system.* The helicopter deck should be protected by CO_2 extinguishers of a total capacity of not less than 18 kg (40 lbs) or

equivalent. One of these extinguishers was equipped so to enable it to reach the engine area of any helicopter using the helicopter deck. The CO_2 extinguishers are to be located so that the equipment would not be vulnerable to the same damage as the dry powder extinguishers.

o *Fixed foam system.* A fixed foam fire extinguishing system, consisting of monitors or hose streams or both, should be installed to protect the helicopter landing area. The locations of the monitors should be reviewed and compliance with the approved system drawings verified.

o *Firefighter's outfits.* In addition to any firefighter's outfits required elsewhere in the Class rules, two additional firefighter's outfits are to be provided and stored near the helicopter deck. These are to be easily accessible in an emergency.

- *Additional equipment.* The following equipment should be provided near the helicopter deck and should be stored in a manner that provides for immediate use and protection from the elements:
  o Adjustable wrench.
  o Fire-resistant blanket.
  o Bolt cutter, with arm length of 60 cm (24 in) or more.
  o Grab hook or salving hook.
  o Heavy duty hack saw, complete with a minimum of six spare blades.
  o Ladder.
  o Lifeline of 5 mm (0.19 in) diameter × 15 m (50 ft) in length.
  o Side cutting pliers.
  o Set of assorted screwdrivers.
  o Harness knife, complete with sheath.
- *Helicopter Facilities.* Enclosed helicopter hangars, refuelling and maintenance facilities are to be provided with fixed fire extinguishing systems. Appropriate details, indicating the coverage of the space, as well as the locations of the controls and extinguishing medium storage should be clearly indicated on the FCP.

Vessels with automation designations

In addition to the other required fire detection and alarm system arrangements stated herein, for vessels which have shipboard automated systems (indicated by an ACC notation or similar), the propulsion machinery space is required to have protection by a fixed fire detection and alarm system. Details indicating compliance with the specific Class rules should be indicated on the FCP and thereafter verified during Class review.

Firefighting station

Vessels being classed with the 'ACCU' automation notation or similar are required to be provided with a 'Firefighting Station'. The 'Firefighting Station' must be provided with certain additional controls and equipment. The location of the 'Firefighting Station', as well as controls and equipment associated with it, should be indicated on the FCP, indicating compliance with the following:

- Location
  The firefighting station must be located outside the propulsion machinery space. However, consideration can be given to the installation of the firefighting control station within the compartment housing the centralised control station, provided that the compartment's boundary common with the propulsion machinery space, including glass windows and doors, is insulated to A-60 standard and the doors opening into the propulsion machinery space are self-closing. In addition, there should be some form of protected access, insulated to A-60 standard, from the room to the open deck. These details should be indicated on the FCP or verified from other drawings.
- Firefighting controls.
  The firefighting station is required to be provided with remote manual controls for the operations identified in the following list and the FCP should clearly indicate these accordingly:
  o Shut down of ventilation fans serving the machinery space.
  o Shut down of fuel oil, lubricating oil, and thermal oil system pumps.
  o Shut down of forced and induced draft blowers of boilers, inert gas generators and incinerators, and of auxiliary blowers of propulsion diesel engines.
  o Closing of propulsion machinery space fuel oil tanks suction valves. This is to include other forms of fuel supply, such as the gas supply valves on LNG carriers.
  o Shut down of fixed local application firefighting systems, before activation of a high-expansion foam fire extinguishing system, to avoid adverse water action on the foam.
  o Closing of propulsion machinery space skylights, openings in funnels, ventilator dampers, and other openings.
  o Closing of propulsion machinery space watertight and fire-resistant doors. Self-closing doors with no hold back arrangements may be excluded.
  o Starting of emergency generator where it is not arranged for automatic starting.
  o Starting of a fire pump located outside the propulsion machinery space, including operation of all necessary valves, to pressurise the

fire main. Starting of one of the main fire pumps is also to be provided on the navigational bridge.

o Actuation of the fixed fire extinguishing system for the propulsion machinery space.

- Fire detection and alarm system.
  In addition to other required fire detection and alarm system arrangements, for vessels which are to receive the ACCU notation, or similar, the propulsion machinery space is required to be protected by a fixed fire detection and alarm system. The FCP should provide sufficient details to indicate compliance with the following:

  o Unattended propulsion machinery space is protected by a fire detection and alarm system.

  o Indicator panel is located at either the navigation bridge or the fire control station.

  o If the indicator panel is located at the fire control station, then a repeater panel is provided on the navigation bridge.

- Fire alarm call points.
  Manually operated fire alarm call points are to be provided at the centralised control station, on the navigation bridge, and in the passageways leading to the propulsion machinery space, and these items must be clearly indicated on the FCP.

- Portable fire extinguishers.
  For vessels receiving the ACCU notation, or similar, Class requirements are such that the vessel must be provided with additional portable fire extinguishers. The number of additional portable extinguishers is to be equal to the number of hand portable extinguishers required for the machinery spaces. These additional extinguishers are to be provided either at the firefighting station or at the entrance to the propulsion machinery space and are to be indicated on the FCP.

VESSELS CARRYING IMDG CLASSED 'DANGEROUS GOODS'

In addition to the requirements discussed throughout this chapter, a vessel designated to carry 'Dangerous Goods', in accordance with the IMDG Code, is required to be fitted with certain additional systems and equipment, depending on the class of Dangerous Goods to be carried. The additional safety and firefighting arrangements are identified in the appendices, and the specific requirements vary depending upon the class or classes of dangerous goods to be carried. Where any additional requirements are set by Class due to the carriage of Dangerous Goods, such arrangements or equipment are required to be indicated on the FCP and the review should verify compliance with the Class rules accordingly.

VESSELS INTENDED TO CARRY OIL IN BULK

In addition to the requirements discussed throughout this chapter, vessels carrying oil in bulk present certain additional fire hazards due in part to the flammability of the cargo, and therefore, certain additional requirements, as discussed later, are required. The review of the FCP should verify these systems are properly indicated and are consistent with the Class-approved system arrangements.

Fixed fire extinguishing systems

Vessels intended to carry oil in bulk are required to carry the following additional fixed fire extinguishing systems:

- A deck foam system covering the cargo deck area is typically required to be installed. The specific arrangements required are discussed in Chapter 6. The FCP should be reviewed to verify that the foam system control station, monitors, applicators, foam main isolation valves, etc., are indicated and that the locations of this equipment comply with the details of the deck foam system approved for the vessel.
- The cargo pump room is required to be provided with a fixed fire extinguishing system, and the controls and arrangements for the system should be clearly indicated on the FCP.

Fire main system

Isolation valves are required to be fitted in the fire main at the poop front in a protected position and on the tank deck at intervals of not more than 40 m (131 ft). Refer to Chapter 3 regarding the IMO interpretation. The arrangements indicated on the FCP should indicate the isolation valves, which will be verified during the Class plan review.

Firefighter's outfits

In addition to the two firefighter's outfits normally required for any cargo vessel, Class rules requires that vessels intended to carry oil in bulk are provided with two additional firefighter's outfits, with the location of these clearly indicated on the FCP.

VESSELS INTENDED TO CARRY LIQUEFIED GAS IN BULK

Like vessels which ship oil in bulk (i.e. oil tankers) gas carriers also present additional fire hazards due to the flammability of the cargo being transported. Accordingly, in addition to the requirements discussed throughout

Chapter 10, gas carriers are also required to be provided with certain additional firefighting systems, which are discussed later. The review of the FCP should verify whether these requirements are properly indicated on the FCP and are consistent with the Class-approved system arrangements.

Additional required fixed systems

Gas carriers transporting flammable or toxic products are required to be provided with a water spray system. Refer to Chapter 7 for the areas to be protected. If a water spray system is required, the indicated arrangements should be consistent with the system details approved for the vessel, and review of the FCP should verify these systems. Where the gas carrier is required to transport flammable products, a fixed dry chemical powder-type extinguishing system for the deck in the cargo area, and the bow or stern cargo handling areas are required. The FCP should clearly indicate the location of these systems. If required, the indicated arrangements should be consistent with the system details approved for the vessel, and review of the FCP should verify these systems comply with Class requirements.

Firefighter's outfits

Every gas carrier that transports flammable products must be provided with the following sets of firefighter's outfits:

- For vessels of 5,000 m^3 and below :   a   minimum   of   four outfits.
- For vessels of 5,000 m^3 and above :   a minimum of five outfits.

In addition, any breathing apparatus required as part of a firefighter's outfit should be of a self-contained air-breathing type having a capacity of at least 1,200 l (317 gal) of free air. If these additional firefighter's outfits are required, the provisions for these should be clearly indicated on the FCP and verified accordingly during the Class plan review.

Gas detection systems

Depending on the cargo to be carried, a gas detection system may be required, in accordance with the Class rules. Where required, the control station and locations monitored are typically indicated on the FCP, and if indicated, the arrangements should be reviewed to confirm that they are consistent with the arrangements approved for the vessel.

Personal protective protection and the provision of first aid equipment

Whilst not specifically associated with firefighting, many Class rules discuss the provision of certain personal protective equipment for personnel and the first aid equipment required to be carried onboard. Whether this is discussed in the Class rules depends largely on the type of cargoes to be carried. This equipment is frequently indicated on the SSP and Ship's Safety Management System. Compliance with the Class requirements should be verified if indicated.

Fire main system

Isolation valves are required to be fitted in the fire main at the poop front in a protected position and on the tank deck at intervals of not more than 40 m (131 ft). Whilst not specifically mentioned in most Class rules, the arrangements indicated on the FCP should indicate the position of the isolation valves, with these verified during the Class plan review.

VESSELS INTENDED TO CARRY CHEMICALS IN BULK

Chemical carriers may present additional hazards depending upon the cargoes to be transported. Accordingly, in addition to the requirements discussed throughout Chapter 10, chemical carriers are also required to be provided with certain additional fire and safety systems, as discussed later. The review of the FCP should verify the systems carried onboard are properly indicated and are consistent with the Class approved system arrangements.

Fixed fire extinguishing systems

In addition to the fixed fire extinguishing systems discussed earlier, vessels intended to carry chemicals in bulk are to be provided with the following additional fixed fire extinguishing systems:

- Vessels intended to carry chemicals in bulk are typically required to be fitted with a deck foam system covering the cargo deck area, as well as any bow or stern loading areas. The specific requirements and arrangements required are discussed in Chapter 6. The FCP should be reviewed to verify that the foam system control station, monitors, applicators, foam main isolation valves, etc., are indicated and that the locations of this equipment comply with the details of the deck foam system approved for the vessel, as necessary.

- The cargo pump room is required to be provided with a fixed fire extinguishing system, and the controls and arrangements for the system should be clearly indicated on the FCP.

Vapour detection system

Depending on the cargo to be carried, the means for flammable and/or toxic vapour detection may be required. Where it is required, such equipment is typically indicated on the SSP, and if portable devices are to be used, the SSP may be the only place where such equipment is indicated. Accordingly, if required for the vessel, the arrangements should be reviewed to conform with Class rules.

Personal protective equipment and safety equipment

While not specifically associated with firefighting, it is common for Class rules to discuss certain protective and safety equipment for personnel, depending on the cargoes to be carried. This equipment is frequently indicated on the FCP or the SSP. Compliance with the Class requirements should be verified, where indicated.

Fire main system

Isolation valves are required to be fitted in the fire main at the poop front in a protected position and on the tank deck at intervals of not more than 40 m (131 ft). The arrangements indicated on the FCP should indicate the location of the isolation valves and verified during the plan review.

PASSENGER VESSELS

Passenger vessels present unique dangers and safety concerns. This is due to the large number of untrained people onboard who will not be familiar with the shipboard emergency procedures and may in certain cases have reduced agility and mobility. Accordingly, there are certain fire and safety requirements that are uniquely applicable to passenger vessels. These requirements are in addition to or in lieu of the requirements discussed throughout Chapter 10 and should be verified during review of the FCP.

Fire main system

The number and location of the main fire pumps are to be indicated on the FCP and indicate that the location(s), as well as the separation and access arrangements, complies with the Class requirements for that vessel. The number and position of hydrants are to comply with the Class requirements for that vessel and must be consistent with the arrangements indicated on the approved fire main system diagram. In this respect, if the FCP does not indicate the maximum coverage areas possible from each hydrant, the verification of compliance will necessarily be left to the discretion of the attending Class Surveyor. However, the locations of the hydrants should still be reviewed, and any questionable areas brought to the attention of the Class Surveyor. Typically, Class rules require that at least one fire hose is provided for each required hydrant. For vessels carrying more than 36 passengers, each interior hydrant is to have a hose attached. Category 'A' machinery spaces on passenger vessels are to be provided with at least two water fog applicators.

Additional fixed fire extinguishing systems required on passenger vessels

Depending on the construction methods utilised, vessels carrying 36 passengers or less may be required to provide an automatic sprinkler system (or equivalent) for the accommodation, controls, and service spaces. Moreover, all passenger vessels carrying 36 passengers, or more, are required to have automatic sprinkler systems installed for such spaces. Separate rules generally apply for cabin balconies. The FCP should indicate the spaces being protected, and the review of the FCP should verify consistency with the approved details of the sprinkler system plans and confirm that all required spaces are accordingly protected.

Fire detection and fire alarm systems

In addition to any previously discussed fire detection and alarm system component location requirements, the location of the fire detectors, fire alarms, manual call points, indicating stations, and control stations required by Class are to be clearly shown on the FCP and verified during review.

Firefighter's outfits

In addition to the two firefighter's outfits required by, every passenger vessel must be provided with two additional firefighter's outfits and two additional sets of personal protective equipment at every 80 m (263 ft) interval, or part thereof, of the aggregate of the lengths of all passenger spaces and service spaces, or if there is more than one such deck, on the deck which has the largest aggregate of such lengths. Each set of personal equipment should comprise non-conducting boots and gloves, a rigid helmet, and protective clothing. For vessels carrying 36 passengers, or more, two additional firefighter's outfits are required to be provided for each main vertical zone. No additional firefighter's outfits are required, however, for stairway enclosures, which constitute individual main vertical zones and for the main vertical zones in the fore or aft end of a vessel that do not contain spaces of the following categories, as defined in SOLAS regulation II-2/9.2.2.3.2.2. These are accordingly:

- Accommodation spaces of minor fire risk.
- Accommodation spaces of moderate fire risk system.
- Accommodation spaces of greater fire risk.
- Machinery spaces and main galleys.

There is also to be provided for each pair of breathing apparatus one water fog applicator which should be stored adjacent to such apparatus.

Storage arrangements

The firefighter's outfits are to be stored so to be easily accessible and ready for use. They are also to be stored in widely separate locations, with at least two firefighter's outfits and one set of personal protective equipment to be available at any one position. For vessels carrying more than 36 passengers, at least two firefighter's outfits are to be stored in each main vertical zone. Also, vessels carrying more than 36 passengers are to carry at least two spare charges for each breathing apparatus in lieu of the one spare charge required by most Class rules, and all air cylinders for breathing apparatus are to be interchangeable.

Additional marking requirements

In conducting the review of the FCP, the engineer should note that SOLAS regulation II-2/13.3.2.5.1 requires all fire equipment location markings to be of some approved form of photoluminescent material or marked by lighting. Compliance with this requirement should be verified during the plan review as a classification requirement.

RORO SPACES

Vessels intended to carry vehicles present certain additional hazards due to the low flash point fuels normally used and the increased amount of class 'A' combustibles. Accordingly, certain requirements for cargo spaces intended for the carriage of vehicles should be verified in addition to those specified throughout this chapter. In addition, the requirements outlined in this chapter would also be applicable if the RORO spaces are on a passenger vessel.

Fire detection

RORO spaces are required to be provided with a fixed fire detection and fire alarm system complying with Class stipulated requirements. The locations of the fire detection and alarm system components, including the locations of the fire detectors, fire alarms, manual call points, indicating stations, and control stations, should be clearly indicated on the FCP and verified during Class review.

Fixed fire extinguishing systems

In addition, the location of the system release controls, as well as storage arrangements for the fire extinguishing medium, should be verified.

RORO spaces capable of being sealed

RORO cargo spaces capable of being sealed must be fitted with a fixed gas fire extinguishing system. As an alternative, Class may allow the installation of a fixed pressure water spray system for such spaces. Indication that such systems are provided, and appropriate details regarding the locations of the control valves and actuating station(s) should be indicated on the FCP and their compliance with the appropriate requirements verified during Class plan review.

RORO spaces not capable of being sealed

RORO cargo spaces not capable of being sealed are to be fitted with a fixed pressure water spray system. Indication that such systems are provided and appropriate details regarding the locations of the control valves and actuating station(s) should be indicated on the FCP and their compliance with the appropriate requirements verified during Class plan review.

Fire hydrants

Any RORO cargo space or any vehicle spaces are to be provided with sufficient hydrants such that at least two jets of water, each from a single length of hose are capable of reaching any part of the space and that hydrants are to be positioned near the accesses to the spaces.

Portable extinguishers

At least one portable extinguisher should be provided at each access to any RORO cargo space. In addition, at each vehicle deck level where vehicles with fuel in their tanks are carried, sufficient portable extinguishers suitable for fighting oil fires are to be provided such that they are not more than 20 m (65 ft) apart on both sides of the vessel. The one located at the access point may be credited for this purpose. These extinguishers should be indicated on the FCP and verified during Class plan review.

Water fog and foam applicators

Each RORO cargo space intended for the carriage of motor vehicles with fuel in their tanks for their own propulsion must be provided with the following:

- At least three water fog applicators.
- One portable foam applicator unit provided that at least two such units are available in the vessel for use in such RORO cargo spaces.

These items should be clearly indicated on the FCP and verified during Class plan review.

CARGO SPACES, OTHER THAN RORO CARGO SPACES, INTENDED TO CARRY VEHICLES WITH FUEL IN THEIR TANKS

Cargo spaces, other than RORO cargo spaces, intended for the carriage of motor vehicles with fuel in their tanks for their own propulsion must comply with Class-mandated provisions except where in lieu a sample extraction smoke detection system complying with the relevant Class provisions may be permitted. Accordingly, the guidance provided in this chapter would be applicable with the exception that a sample extraction smoke detection system may be installed in lieu of the fire detection system. Details verifying that the appropriate systems and equipment are provided for such spaces should be indicated on the FCP and verified during the Class plan review in association with the review of requirements outlined throughout this chapter. As per SOLAS regulation II-2/20–1, when a vehicle carrier carries as cargo one or more motor vehicles with either compressed hydrogen or compressed natural gas in their tanks for their own propulsion, at least two portable gas detectors should be provided. Such detectors must be suitable for the detection of the gas fuel and be of a certified safe type for use in the explosive gas and air mixture.

VESSELS UNDER 500 GRT

For vessels under 500 GRT, certain alternative arrangements may be permitted by Class. Where the FCPs for such vessels are under review, the alternative requirements, as discussed later, should be considered in association with the requirements stated throughout this chapter, as applicable.

Fire pumps

The Class rules for each specific type of vessel under 500 GRT will address the number and type of fire pumps required. Separation arrangements for the fire pumps as referenced throughout this chapter are therefore not applicable.

Machinery space fixed fire extinguishing systems

Fixed fire extinguishing systems are not required for machinery spaces of vessels below 500 GRT, unless a space contains propulsion and auxiliary engines with a total aggregate power of 750 kW (1,000 bhp) or greater or an oil fuel unit for heated fuel oil, regardless of the total aggregate power.

Carbon dioxide (CO_2) systems

Alternative storage arrangements for the CO_2 system are usually permitted, in accordance with individual Class rules.

Fire axe

A fire axe is required to be carried on all such vessels over 20 m (65 ft) in length. Firefighter's outfits are not required.

Oil tankers

The cargo tank protection for oil carriers of less than 30.5 m (100 ft) in length may be provided with two 'B-V' extinguishers in place of a deck foam system.

In this final chapter, we have examined the role and function of the FCP, as it relates to the universally accepted Class rules. The FCP is a critical document or set of documents and forms one of the key elements of the SSP, which in turn is a critical element of the Ship's Safety Management System. It is vitally important that the FCP is updated and reviewed regularly – both internally and externally by Class – to ensure that it meets the latest standards and complies with the most current Class rules and requirements.

Class reviews, as discussed throughout this book, are an essential method of ensuring the merchant fleet meets the latest standards in shipboard firefighting. When emergencies happen at sea, there is minimal scope for error. Fire has always been, and remains, one of the worst types of emergencies that can occur on any oceangoing vessel. When appropriate measures are not taken, this puts lives invariably, and unnecessarily, at risk. Shipboard fires can also cause inestimable damage to the marine environment. It is therefore everyone's responsibility to ensure shipboard processes, systems, and infrastructure meet and comply with Class requirements.

References

The following documents and publications have been referenced or utilised as sources of information in the development of this book:

ABS Guide for building and classing accommodation barges (2014).

ABS Guide for building and classing yachts (2014).

ABS Rules for building and classing high speed craft (2014).

ABS Rules for building and classing steel barges (2014).

ABS Rules for building and classing steel vessels (2014).

ABS Rules for building and classing steel vessels for service on rivers and intra-coastal waterways (2014).

ABS Rules for building and classing steel vessels under 90 m (295 ft) in length (2014).

"A firefighter's guide to foam," National Foam Company.

Alternate compliance programme (US supplement to the RINA rules for the classification of ships) 2011.

Ansul Marine Foam Systems Design, Installation and Maintenance Manual.

"Fire Protection Handbook," 19th Edition, National Fire Protection Association.

IMO MSC Circular 847, Interpretations of vague expressions and other vague wording in SOLAS chapter II-2, Consolidated Edition, 1997.

IMO Resolution A.602(15), Revised guidelines for marine portable fire extinguishers.

International Association of Classification Societies (IACS) Unified Interpretations.

International Maritime Organisation (IMO) Safety of Life at Sea (SOLAS), Amendments 2000, Published in 2001.

International Maritime Organisation International code for fire safety systems (FSS Code), Published in 2001.

Lloyds Register Guidance notes for risk-based analysis: fire loads and protection 2014.

LY3 The large commercial yacht code 2012.

"Marine fire prevention, firefighting and fire safety," Marine Training Advisory Board, US Department of Transportation.

MGN 258 (M+F) Decommissioning of Halon systems.

MGN 354 (M+F) Fishing and small vessels – safe operation of fixed CO_2 gas fire extinguishing systems.

MGN 389 (M+F) Operating instructions and signage for fixed gas fire-extinguishing systems.

MSN (M) 1359 Emergency equipment lockers for RORO passenger ships 1988.

MSN 1646 (M) The merchant shipping (lifesaving appliances for passenger ships of classes III to VI(A)) regulations 1999.

MSN 1646 The merchant shipping (lifesaving appliances for passenger ships of classes III to VI(A)) (amendment) regulations 2001.

MSN 1665 (M) The merchant shipping (fire protection) regulations 1998: fire-fighting equipment.

MSN 1666 (M) The merchant shipping (fire protection) regulations 1998: fixed fire detection alarm and extinguishing systems.

MSN 1670 (M) The merchant shipping (fire protection) regulations 1998: exemptions.

MSN 1747 (M) The merchant shipping (passenger ships on domestic voyages) regulations 2000.

SFPE Handbook of fire protection engineering, 3rd edition, national fire protection association.

US Coast Guard NVIC 6–72, Guide to fixed firefighting equipment aboard merchant vessels.

Index